The Ethics of Agribusiness

This book offers an original perspective on food supply chains. It argues that the ability to trade food on a global scale *could be* intrinsically good aside from any instrumental value that people gain from it.

While the author's argument seems to counter wholesale anti-agribusiness views, it is consistent with the larger goals of food-justice movements. The author examines the structures of food supply chains, revealing the kinds of harm they help produce. They include slavery, abusive labor, geopolitical exploitation, ecological degradation, and public health impacts. Although the book argues that food supply chains can be collectively beneficial, eliminating their immoral features must hold steady as a continuous enterprise. Securing this outcome means that we go beyond critique. The final chapter advocates for the sustainable food label to address issues of food justice and food sovereignty.

The Ethics of Agribusiness will interest researchers and advanced students working in food ethics, environmental ethics, and agricultural ethics.

Shane Epting is an Assistant Professor of Philosophy at the Missouri University of Science and Technology. His previous books include *The Morality of Urban Mobility: Technology and Philosophy of the City* and *Saving Cities: A Taxonomy of Urban Technologies*.

Routledge Research in Applied Ethics

The Ethics of War and the Force of Law
A Modern Just War Theory
Uwe Steinhoff

Sexual Ethics in a Secular Age
Is There Still a Virtue of Chastity?
Edited by Eric J. Silverman

The Ethics of Virutal and Augmented Reality
Building Worlds
Erick Jose Ramirez

The Philosophy of Online Manipulation
Edited by Fleur Jongepier & Michael Klenk

The Ethics of Political Dissent
Tony Milligan

Why Conscience Matters
A Defence of Conscientious Objection in Healthcare
Xavier Symons

The Ethics of Agribusiness
Justice and Global Food in Focus
Shane Epting

What Kind of Death
The Ethics of Determining One's Own Death
Govert den Hartogh

For more information about this series, please visit: https://www.routledge.com/Routledge-Research-in-Applied-Ethics/book-series/RRAES

The Ethics of Agribusiness

Justice and Global Food in Focus

Shane Epting

Routledge
Taylor & Francis Group
NEW YORK AND LONDON

First published 2023
by Routledge
605 Third Avenue, New York, NY 10158

and by Routledge
4 Park Square, Milton Park, Abingdon, Oxon OX14 4RN

Routledge is an imprint of the Taylor & Francis Group, an informa business

Library of Congress Cataloging-in-Publication Data
A catalog record for this title has been requested

ISBN: 978-1-032-18570-5 (hbk)
ISBN: 978-1-032-18638-2 (pbk)
ISBN: 978-1-003-25550-5 (ebk)

DOI: 10.4324/9781003255505

Typeset in Sabon
by Taylor & Francis Books

This book is dedicated to the memory of Jarred Reynolds

Contents

Acknowledgments

I appreciate the philosophers and thinkers whose works were helpful in thinking through pressing issues raised in this book. Robyn Metcalfe is the first person that comes to mind: *Food Routes* was a pleasure. Julie Guthman's *Weighing In* was inspiring and enlightening. Thanks to the (mostly former) Michigan State Philosophy of Food crew: Paul Thompson, Samantha Noll, Zach Piso, and Ian Werkheiser. Thanks to The Culinary Mind people: Andrea Borghini, Patrick Engisch, and Nicola Piras. Also, thanks to David M. Kaplan and Joey Tuminello. Thanks to the Interdisciplinary Environmental Association. Thanks to the Philosophy of the City Executive Committee (past and present): Michael Menser, Jules Simon, Ronald Sundstrom, Michael Nagenborg, and Sanna Lehtinen. Thanks to Mark Bartos, Andy Haight, Shelley Rios, Carl Sachs, and Warren Stoltzfus. Of course, special thanks to Ana and her family.

Most of my article, "Unjust food systems and applied mereology," published in *Argumenta* (2020, Vol. 10, No.1, pp. 199–213) appears in revised form as parts of Chapter 3 and Chapter 4. A few revised paragraphs in Chapter 1 first appeared in my article, "Participatory budgeting and vertical agriculture: A thought experiment in food system reform," in the *Journal of Agricultural and Environmental Ethics* (2016, Vol. 29, No. 5, pp. 737–748).

Preface

In Plato's *Apology* (1966, 38a) Socrates' line goes: "the unexamined life is not worth living," but I would quip that "the unexamined dish is not worth eating." Experience has taught me that investigating food's backstory rarely enhances it unless you eat at a farm-to-table joint. Even then, though, they cannot always tell you if the workers were paid fairly and had health coverage. This notion is a glimpse of what is to come in this book.

Learning about contemporary slavery in agribusiness, farmworkers' unethical treatment, factory farming ills, and widespread ecological degradation induces anger and depression in most people. Yet, the forces in our lives make it practically impossible to live without intersecting with such harm unknowingly. How can we come to grips with several unfortunate realities? I answer this question by keeping the future in view—with hope and strategy. Getting there is the motivation driving this research.

Although this book deals with the interdisciplinary topic of agribusiness, its orientation remains philosophical. It takes the ordinary and examines it in a way that reveals what is extraordinary. In this case, the "ordinary" is food supply chains, or just "food chains," as I call them. The "extraordinary" is uncovering what is beyond appearance, data, and narrative. One could call it mining complex notions from the everyday world.

In turn, food scholars have much to gain from this text if they have the patience for an enterprise that goes beneath the facts, figures, and stories associated with traditional pursuits. Philosophers, at least mainstream ones, can see how tools of the trade can help make sense of flesh-and-blood experiences. Philosophers of food and similar mavericks who have no problems crossing disciplinary boundaries might feel at home reading these chapters. Making progress that involves a comprehensive view of agribusiness requires leaving disciplinary comfort zones. In that sense, this book serves as a tour guide for daring explorers willing to examine the moral complexity of getting food from seeds to stores to stomachs.

I wrote this book not to alienate any of the above groups or put them to sleep. It should be accessible to anyone who wants to think through agribusiness issues. While it was tempting to delve into long-standing philosophical disputes, many of those pursuits were sidestepped, acknowledged in notes, and still lurk in the background in some instances. The aim was to stay on track, and the sophisticated nature of food problems requires attention. While the issues covered here exist for scholarly purposes, they concern real people. Compared to food-justice activists, these efforts are incredibly modest at best.

It will take concerted measures to move toward reforms and progress. Advances could come from several sources, including but not limited to communities, municipalities, improved policies, and agricultural technologies. Others can come from "technologies of the mind," well-ordered reasoning with precise objectives. It will take numerous people who care about these issues to improve them, philosophers included.

Reference

Plato. *Plato in Twelve Volumes, Vol. 1*, translated by Harold North Fowler; Introduction by W.R.M. Lamb. Cambridge, MA: Harvard University Press; London, William Heinemann Ltd. 1966.

1 The First Link

Introduction, Methodology, and Overview

Introduction

Philosophers of food have made tremendous contributions to a growing body of literature in the past few decades, with recent additions showing the numerous distinct research strands in this subfield. Here are some examples of these fascinating works. A prolific philosopher researching agriculture, Paul Thompson (2010, 2007) deals with numerous topics in this area, from agrarianism in contemporary agriculture to the ethical dimensions of biotech foods and beyond. Carolyn Korsmeyer (1999) thoroughly explores the experience of taste. Andrea Borghini (2015) writes on the ontology of recipes and authenticity in foods. Nicola Perullo (2012, 2020) philosophically explores wine, extensively. Samantha Noll (2015, 2017) researches urban agriculture. Ian Werkheiser (2018) examines duties to livestock. Together, they have written about local food movements (Werkheiser and Noll 2014).

The specific place in the literature where this book would fit is in response to Werkheiser and Noll's insights. Their recent article, "From food justice to a tool of the status quo: Three sub-movements within local food," has garnered much attention in philosophy of food and beyond. They identify the strengths and weaknesses of reform efforts, calling them individual-focused (IF), system-focused (SF), and community-focused (CF). IFs rely on each person's lifestyle choices. The goal is for the well-being of the individual, which includes people shopping at local farmers' markets and contributing to the economic viability of their community, gaining a sense of civic responsibility. Striving to be environmentally considerate, individuals reduce the miles that some foods must travel based on the assumption that such actions are effective. Michael Pollen's (2009) *A defense of food: An eater's manifesto* is an exemplar, holding that the long-term impacts of such choices will have profound effects.

While IF approaches show how the person can become an ethical consumer, this position receives criticisms about its feasibility. For instance, one could argue that it is doubtful that relying on individuals' lifestyle changes to improve food systems through shopping habits can bring the wide-scale shift that is required to improve agribusiness (Werkheiser and Noll 2014).

DOI: 10.4324/9781003255505-1

Regarding relying on locally grown produce to reduce unwanted ecological impacts, critics of food miles such as Justin McWilliams (2009) exhibit how such calculations fail to consider energy concerns that determine efficiency. He cites instances showing that it is environmentally sound for specific cities to import certain crops rather than locally growing them.

Calling for greater certainty than IFs, SF sub-movements suggest that agricultural issues are too deep-seated for IF solutions, requiring policy decisions to combat the harmful effects of agribusiness (Werkheiser and Noll 2014). While IFs urge individuals to support farmers' markets, SFs endorse policies and organizations that can restructure food systems so that sustainable, just institutions become the rule, not the exception. Even though the ambitions of SFs seem as if they could change food systems due to their enforceable policies and laws, they do not consider the more extensive globalized condition of food production. For instance, it is somewhat naïve to think that municipal ordinances can defeat multinational food conglomerates' lobbying powers and deep pockets. This outlook does not dismiss the possible power of gradual changes to food policies but acknowledges their limitations.

Considering the factors above, Werkheiser and Noll (2014) champion CFs because they shift the food-system conversation away from food security to food sovereignty. Put briefly: the latter involves communities having autonomy over their food system (Navin 2014; Grey and Patel 2015). While IFs and SFs view people and food as separate, CFs emphasize interconnections between them (Werkheiser and Noll 2014). Through gaining some control over food production, CFs advance beyond IFs and SFs. While I also champion CFs in spirit and elsewhere (Epting 2016), additional research on SFs could yield insights into how to maximize their effectiveness at reform.

Alternatively, but possible, although it does not exist in the literature, a fourth category, agribusiness-focused (AF) sub-movement, could exist. It would rely on business to reform itself, a laughable measure—unless doing so would increase profits. Still, there is the possibility that AFs could want to participate in reform efforts. Perhaps ethically minded entrepreneurs and professionals working in logistics would wish to spearhead such pursuits. The reality is that agribusiness needs people who would act ethically. I am thinking about the kind of mindsets that people leading food-justice movements have rather than cookie-cutter executives. They could join the IF–SF approach to deliver better food worlds. This book moves in that direction, and it also aims to uncover the benefits of the global food trade that remain out of immediate view. Food connects us in more ways than imaginable at first glance, and investigating such connections has much to teach us.

Exploring the connections between philosophy and food is not a contemporary endeavor, considering that (Sister) Sor Juana Inés de la Cruz wrote that people could learn philosophy in the kitchen over 300 years ago. Aristotle would have had better acumens—if only he had paid attention to the ways that foods change during the cooking process (de la Cruz et al.

2009; Epting 2017). While this process seems ordinary, a proper inquiry and interrogation reveal its extraordinary character. People can understand metaphysical principles by observing events physically in the world. In turn, the kitchen is a fantastic laboratory for thought experiments. When it comes to the current state of today's global food trade, the wisdom underpinning her insights extends from the ontological to the ethical, and they bring numerous areas of study into view.

For instance, examining the ingredient list in most food-like products reveals a litany of ingredients and food-like "things" that were most likely assembled and baked far from their points of origin, then distributed far and wide. The behind-the-scenes story of how a food chain's parts fit together is fascinating. Such tales depict scenes from farms to shipping centers, warehouses, distribution centers, vendors, stores, pantries, and restaurants (Metcalfe 2019). It makes for very entertaining reading. Still, there are many more parts to such puzzles lurking out of view, including but not limited to policies, loans, labor practices, trade agreements, national and state regulations, cultural elements, and politics—just to name a few. Sorting out and classifying how all ingredients travel from farms to store shelves is nearly impossible in the worst sense and a perpetual process in the best-case scenario.

Beyond such a goal, ethical evaluations are needed to make moral sense of such arrangements. Robyn Metcalfe (2019) brings this point to light in her recent narrative-esque book, *Food Routes: Growing Bananas in Iceland and Other Tales from the Logistics of Eating*. While researching for this book, ethical examinations geared explicitly towards food logistics were in short supply. Many of the crucial elements such as labor and working conditions remain hidden, and they would undoubtedly change the focus of food chains' stories if consumers knew such realities. Considering these items in tandem shows that, at minimum, a two-part process is required to help create an ontological-ethical panorama of global food. Fleshing out such a view is this book's purpose. Achieving the desired end means using the best tools for the job. Although I use specific philosophical devices to make sense of agribusiness, I employ lessons from many disciplines across the academy and reports from several fields to bolster its theoretical grounding. In turn, developing this (non-traditional) mixed-methods approach can reveal the essential insights that yield a perspective of the worldwide food exchanges that enhances the ability to support worthwhile goals such as ethically sound sustainable food chains.

On this note, it is worth mentioning that I almost exclusively refer to "food chains" rather than "food supply chains" or "food systems" in this book. Of course, a food chain is a system. The latter somewhat vague term does not illustrate what it does or inform us about its characteristics, saying what kind of a system it concerns. "Food chain," however, says that the given foods are part of a moving system that belongs to agribusiness. Today's food travels extensively from the field to the distribution center to the processing plant and several other stops before it lands in your stomach.

Done right, this reality calls for celebration. However, "done right" requires chapters to unpack.

To begin, here is this book's thesis: global food chains *could be* intrinsically good, but numerous links are bad. "Bad," however, has a variety of meanings in food-chain cases in this book (and beyond). It could mean "unfortunate" on one end of the scale, and it is "entirely immoral, horrid, and repulsive" when we discover that it involves slavery. On one far end, this condition creates a tension that we must ease. On the other end, it exposes situations that every decent human being should be motivated to reform. This enterprise can save food chains' remarkable dimensions while confronting agribusiness's unethical actors and outcomes. In turn, this undertaking bolsters food-reform efforts. This chapter's purpose is to describe the plan of attack and provide a brief overview of the topics that follow. The first step is becoming familiar with the kinds of issues encountered along the way.

The aspect that requires spotlighting is that we ultimately deal with technology as it exists within a business context, hence the term "agribusiness." While many elements such as policies and kinds of thinking might appear as auxiliary or external efforts for food trade globally, this book explores how we can situate them as technologies, a move that provides congruence when analyzing the totality of food chains. The benefit of this positioning is that we gain a panorama of agribusiness, which provides a way to see the connectivity of issues that appear disjointed. Considering that all choices inherently concern ethics, many of the problematic issues found in food chains require identification before morally examining them. The worry rests with the idea that such affairs remain out of sight for most food consumers, denying their agency in many instances. This condition is an outcome of the global food trade, an unfortunate scenario.

The good news is that we can remedy such situations and improve them, creating food chains for celebrating. The even better news is that pinpointing the kinds of issues that we want to eliminate from these chains reveals their possibility to have intrinsic value, which they can gain through reform. What is required for reform efforts to "do it right" is to look at food chains as integral wholes that need better parts to remove or replace bad ones and install new parts that improve food chains' functioning ethically based on the outcomes delivered. The goal is to make this progress in the minds, fields, seas, roads, skies, processing centers, transport lines, distribution centers, store shelves, restaurant menus, and refrigerators that compose food chains. In turn, the section below shows how the above ideas unfold, serving as a primer to familiarize the concepts explored in this book.

Overview

This book is about food chain reform, meaning that we must get a sense of what needs reforming. Even though looking at cases of concern is highly upsetting, saddening, and frustrating, it must be done to gain a picture of what, when, why, and how humankind should be motivated—eager—to pursue reform efforts. Chapter 2 does just that. However, instead of merely listing itemized wrongdoings and morally dubious affairs, it categorizes stakeholder groups that receive various kinds of harm. This move sets the stage for Chapter 3, wherein I show how the pattern of thought behind "applied mereology" can benefit efforts to identify and organize such harms, focusing on part-to-part relations as each "part" basically is a "link" in the food chain.

Studying such arrangements' relationships and effects help identify a unique issue for agribusiness: the problem of globalized opacity, the topic examined in Chapter 4. The central idea is that due to the nature of today's food chains (and) or involving innumerable parts spread across vast distances, the ability to know about the troubling matters covered in Chapter 2 remains incredibly challenging, if not impossible. Developing the means to conceptualize this condition gives a target to turn attention toward to know how to mitigate harms that arise from this condition. The point is not that globalized opacity must be inherently wrong, but existing harms can continue unabated without assessments and corrective measures.

While the processes above appear as worthwhile undertakings in their own rights, viewing them in tandem reveals that the worrisome contemporary character of agricultural technology, as found in agribusiness, challenges an earlier, established taxonomy of technologies. This situation arose due to the reality of climate change as a wicked problem. Chapter 5 brings this issue into focus, providing new productive conceptualization that adds to this taxonomy, "wicked" technology that got humankind into this mess, and "saving" technologies that can help get out of it. With a clear view of the nature of agribusiness's problems and outcomes that reform needs to deliver, the stage is set for dealing with agribusiness ethically in Chapter 6.

Rather than employing a standard theory from the canon of moral thought, I argue that the manner wherein we approach the different kinds of issues identified in Chapter 2 is—itself—a moral problem, the problem of moral prioritization (Epting 2021). This idea holds there is an order concerning whom or what should receive ethical consideration when dealing with an issue involving multi-stakeholder groups. The process employed to guide the thinking here is "moral ordering." Being proactive in this regard means establishing a "moral order of operations" to approach what appears to be an innumerable problem associated with food chains.

An interpretation of this point by example means that despite having countless moral issues associated with agribusiness, acting first for the people most suffering is the priority. For instance, if people are being forced to farm tuna, we should care about them before ensuring that dolphins are safe from the nets. This point does not entail that dolphin lives do not matter. This reasoning is not a case of misdirection. Both of these stakeholders matter in the totality of food chains. However, I cannot look at a situation like that and want to create a policy that puts the dolphins' well-being in front of a people held in slavery. While it is convenient to argue for universals that can remedy all such situations, operating in the real world might not work that way. We might have to make imperfect progress. Prioritization matters. Moral ordering is a tool for dealing with situations involving two or more stakeholders that seem to be at odds. Rather than trying to force the world to abide by this direction of this theoretical device, its suggestive nature maintains room for flexibility, allowing users to shift the order to accommodate the circumstances that arise in the world.

A better view of food chains comes into view by illustrating how reform efforts should move to avoid further ethical complications. One noticeable aspect is that we should be ethically motivated to remedy *all* such situations. On the extreme end, one could hold that efforts should focus on dismantling globalized food chains as much as possible, reducing or eliminating globalized opacity and the wrongs that it shields from consumers. While the idea that we need to work diligently to improve food chains morally is insightful, this notion requires additional exploration.

Getting rid of dependence on food chains or dismantling them misses an opportunity to reveal a monumental dimension about food chains: they could be intrinsically valuable because of what they can accomplish. While the ability to produce, distribute, and consume food worldwide has instrumental value for dealing with matters such as nutrition and essential caloric intake, the ability to do so is a significant achievement of humankind. It is valuable in and of itself. Only the reality that food chains cause significant harm challenges this notion. While food chains accomplish so much, we cannot say that they have intrinsic value if they cause tremendous harm. If we can imagine what food chains would look like without this harm, we can see their intrinsic value.

Yet, even though this notion discovers something extraordinary in the ordinary, revealed with well-ordered reason operating as a technology, it is only the beginning of discovery. Beyond this point lies something else that requires revealing: modern, globalized food chains involve several other entities that also bear intrinsic value, including but not limited to mathematics, scientific comprehension, engineering principles, logistics, and other specialized bodies of knowledge that we assign intrinsic value. This condition, a confluence of intrinsic values that exceed primary intrinsic value, which I call super-intrinsic value, makes sense. In turn, a clearer picture of the significance of maintaining global food chains comes into view, showing their importance for humankind.

As described in the previous chapter, the project then develops mitigatory measures that could help restore moral integrity to food chains. The most likely way to make progress in this regard is to replace and create technologies that play roles in generating harm, as examined in Chapter 2.Chapter 7 provides two examples. One example is physical and already exists, vertical farming, focusing on how it can help produce desired outcomes in select cases. The other is a technology in the form of a policy-backed "sustainable" food label that could provide customers avenues that deliver better food worlds. It is presented as a kind of "thought experiment."

This experiment exhibits the flabbergasting changes required to make food chains "sustainable" according to a definition that entails justice. By revealing this sobering reality, a view comes into focus showing the work required to deliver better food worlds. It shows the daunting nature of the food-justice enterprise. Thought about in this manner, it is challenging to say that food justice is achievable as if it were an attainable goal. This notion is not meant to discourage food justice. Instead, it puts it in a realistic perspective.

The focus shows that food justice functions more accurately as a verb, an action that we must work toward. Not engaging in this activity means settling with business as usual, which is morally unpalatable. Although incredibly disheartening, there is a glimmer of hope resting in possibilities. Spotlighting them is one goal. Providing an ethically manageable way to approach them is another.

Conclusion

This chapter describes how this book unfolds, starting with situating it within the broader philosophy of food literature and specifically within the research strand focusing on agriculture. The exact place is fleshing out how individual-focused and system-focused sub-movements can work in concert to help reform efforts. It also included speculation holding that businesses that take the lead could work towards meaningful reform efforts that align with the ideas explored in the book, creating an agribusiness-focused sub-movement.

Although these notions are worthwhile pursuits, they are not the motivation driving this book. The aim is to reveal why food chains matter in the grander sense for humanity as an approach to value appreciation and moral action. Even though food chains remain subject to numerous ethical criticisms, their underlying structure can have intrinsic value. Once we examine this structure closely, we learn more about it and humanity, which is the additional significance of this book.

References

Borghini, A. (2015). What is a recipe? *Journal of Agricultural and Environmental Ethics*, 28(4), 719–738.

de la Cruz, J. I., Sister, Arenal, E., and Powell, A. (2009). *The answer: Including Sor Filotea's letter and new selected poems*. New York City: The Feminist Press at The City University of New York.

Epting, S. (2016). Participatory budgeting and vertical agriculture: A thought experiment in food system reform. *Journal of Agricultural and Environmental Ethics*, 29 (5), 737–748.

Epting, S. (2017). Introduction to food justice and the built environment. In I. Werkheiser and Z. Piso (Eds.), *Food justice in US and global contexts: Bringing theory and practice together* (pp. 89–93). Dordrecht: Springer.

Epting, S. (2021). *The morality of urban mobility: Technology and philosophy of the city*. London: Rowman & Littlefield International.

Grey, S., & Patel, R. (2015). Food sovereignty as decolonization: Some contributions from indigenous movements to food chain and development politics. *Agriculture and Human Values*, 32, 431–444.

Korsmeyer, C. (1999). *Making sense of taste: Food and philosophy*. Ithaca, NY: Cornell University Press.

Metcalfe, R. (2019). *Food routes: growing bananas in Iceland and other tales from the logistics of eating*. Cambridge, MA: MIT Press.

McWilliams, J. (2009). *Just food: Where locavores get it wrong and how we can truly eat responsibly*. New York: Little, Brown and Company.

Navin, M. (2014). Local food and international ethics. *Journal of Agricultural and Environmental Ethics*, 27, 349–368.

Noll, S. (2015). History lessons: What urban environmental ethics can learn from nineteenth century cities. *Journal of Agricultural and Environmental Ethics*, 28(1), 143–159.

Noll, S. (2017). Food sovereignty in the city: Challenging historical barriers to food justice. In *Food justice in US and global contexts* (pp. 95–111). Cham: Springer.

Perullo, N. (2012). Wineworld: Tasting, making, drinking, being. *Rivista di estetica*, 51, 3–48.

Perullo, N. (2020). *Epistenology: Wine as experience*. New York: Columbia University Press.

Pollen, M. (2008). *A defense of food: An eater's manifesto*. New York: Penguin Press.

Thompson, P. B. (2007). *Food biotechnology in ethical perspective*. Dordrecht: Springer.

Thompson, P. (2010). *The agrarian vision: Sustainability and environmental ethics*. Lexington: University Press of Kentucky.

Werkheiser, I. (2018). Precision livestock farming and farmers' duties to livestock. *Journal of Agricultural and Environmental Ethics*, 31(2), 181–195.

Werkheiser, I., & Noll, S. (2014). From food justice to a tool of the status quo: Three sub- movements within local food. *Journal of Agricultural and Environmental Ethics*, 27(2), 201–210.

2 Food Problems

Introduction

Most available foods depend on numerous technologies for production (Metcalfe 2019). In several cases, food *is* a technology. While these two points suggest that food has become more complex over the last several decades, the moral issues accompanying it have as well. Today, we need different lenses to see how food brings these matters into view. There are several concerns for individual workers, harmful conditions seen at the person's level. These kinds of topics are found across the globe, some with horrendous circumstances. Other individuals, such as consumers, deal with a litany of health issues that arise from harmful sorts of foods that remain in abundance. Yet, sadly, people die of starvation, suffer from malnutrition, and lack practical access to nutritious food.

On a larger scale, global food production plays a significant role in climate change in myriad ways, from methane exhaust from cattle ranching to fossil fuels used to transport goods across oceans, lands, and skies.[1] Although the kinds of concerns above are non-exhaustive, they show a vast range of issues that require attention. The purpose of this chapter is to illustrate the variety of topics that concern farmworkers, consumers, non-government organizations, governments, activists, food scholars, philosophers, and select businesses.

It begins by examining how foods harm individuals, starting with some instances wherein food laborers have drawn attention to numerous issues that cause harm the world over. Next, it investigates how people in countries such as the United States suffer from the outcomes that are seemingly unconnected to the individuals who produce the ingredients that compose the "food-like" goods that they consume. Also, this chapter looks at urban residents who must contend with food insecurity and medical conditions associated with such affairs.

With a firm view of how food impacts persons, the attention turns to how food affects nonhumans, could harm future people, and culturally significant artifacts. These issues intersect with several other matters, including but not limited to public health, urban planning, and international relations. Once

DOI: 10.4324/9781003255505-2

there is a view illustrating how food issues start on the individual level, making their way to the global sphere, the attention turns towards topics such as the role of international food trade in climate change. Gaining a view showing how food affects a wide range of stakeholders remains vital for understanding the impetus behind gearing agribusiness towards more ethical outcomes, which is this transdisciplinary book's purpose. This chapter concludes by setting the stage, moving toward discovering advantageous ways that can eventually help mitigate agribusiness-related harm and transition to more just practices.

The Problematic Positioning of Food

Although humans cannot survive without food, the manner wherein we secure it often goes against the integrity of our survival. The harmful outcomes associated with food production affect every level of being, from the individual to a "planetary collective." From the seed to the table, morally problematic elements permeate every dimension of food production. Yet, none of those aspects are inherent to any part of the process. This point suggests that troubling features are replaceable. The challenge is identifying them and making a case for why they are areas of concern.

Researchers across the academy have made numerous cases that exhibit the countless problems with how the global food trade harms people in the ways mentioned above (Gilson and Kenehan 2018). Arguably, the most significant people who deserve moral attention in the global food supply are workers in the fields, factories, warehouses, ships, hubs, and highways (among others).[2] Without them, the food could not arrive, at least until artificial intelligence (AI) replaces them to lessen costs, which food scholars predict is inevitable (Metcalfe 2019). The places where people deal with food's numerous facets as listed above—along with several additional processing centers, storage facilities, distribution centers, stores, and restaurants—make up food chains.

Each entity involves people who make their living by helping food move from one place to the next. In turn, each stop qualifies as an opportunity for ethical assessment. Although various cities, states, and nations have labor laws, not all such laws and regulations are sufficient to safeguard workers (e.g., Arcury et al. 2002; London 2003; Schwarz et al. 2015). Still, simply because a practice is legal or protected does not entail that they are also ethical. Bearing this point in mind, the following section begins that journey of identifying the issues of concern that will form the backdrop to the ontological, ethical, and mitigatory explorations that arrive in subsequent chapters.

Seeds of Moral Trouble

Despite numerous sources that we use to pinpoint such problems that come from scholarly outputs, organized farmers and food workers worldwide detail the scope of agribusiness's contentious conditions. They have organized and called for, essentially, a radical restructuring of agribusiness when examining the extent of their concerns. For instance, the manner they conceive of themselves, leadership, and representation challenge traditional thought patterns that deal with the ideals behind those terms. This group, La Via Campesina (n.d. para. 10) strives, a "movement of movements" as they refer to themselves, makes this notion apparent. For example, in defining themselves, La Via Campesina (n.d. para. 1) holds:

> La Via Campesina, founded in 1993, is an international movement bringing together millions of peasants, landless workers, indigenous people, pastoralists, fishers, migrant farmworkers, small and medium-size farmers, rural women and peasant youth from around the world... . La Via Campesina comprises 182 local and national organizations in 81 countries from Africa, Asia, Europe and the Americas. Altogether it represents about 200 million small-scale food producers.

The passage above seems like this organization resembles any other group that allies to advance goals that appeal to shared interests. In that sense, La Via Campesina does take the shape of numerous organizations that operate without borders. This familiar form gives the group the ability to have power, putting them on par with other international associations of the same ilk. However, when examining the internal structure and operational procedures they employ to achieve their goals, one sees that they drastically differ from most entities with a readily recognizable hierarchy. While this point might appear insignificant at first glance, a deeper inspection reveals that it gets at the thought structures and habits inextricably bound to the power structures that control food chains worldwide. For instance, La Via Campesina (n.d. para. 8) maintains:

> La Via Campesina is a grassroots mass movement whose vitality and legitimacy comes from peasants' organizations at the grassroots. Its operation relies on decentralized decision-making between all its regional arms... . To ensure that the International Movement represents the diversity of its members, the international operative secretariat of La Via Campesina "rotates" according to the collective decision made every four years by the International Conference.

As shown in the passage above, the organization and operational structure hold importance because it does not see power in the same manner as it exists in companies, non-governmental organizations (NGOs), or some governments. It differs so significantly that "power" might not even be the proper term to describe the kind of force or thinking that they have to employ. This point aside, what we need to stay focused on is that they are not merely striving to build an association that can beat global food at the game they created and controlled. Instead, they are responding to the conditions that emerge from global agribusiness. This notion is also visible in how they situate representation as a concept capable of utilizing collective interests and individual autonomy.

For instance, La Via Campesina (n.d. para. 10) strives to conceptualize representation on the terms that they put forth:

> La Via Campesina is a Movement of Movements. It is not a Federation or an NGO with rigid structures, staffing and funding. As a movement, it articulates the voice of peasants and defends their rights in spaces at all levels… . It ensures direct representation of rural communities and their representatives in global, regional and national spaces of governance and negotiations. La Via Campesina is not speaking "on behalf of" peasants. The peasants that form this global movement articulate directly at the international level.

This passage cements the notion that La Via Campesina exists as an entity that not only pushes against the status quo. Instead, they are primarily removed from the very practices that guide the existence of the power structures that significantly influence all-things food, from the farm to the table. Yet, we still do not have a clear picture of what conditions drive small-scale farmers and workers to want to change the system. One can infer that unfavorable conditions motivate them. This inference is correct. In turn, the following section examines a broad range of issues that trouble food workers in numerous areas of agribusiness. While these conditions are not mentioned explicitly by La Via Campesina, they speak to the spirit of their concerns. These factors bring the desire to control food into view, requiring us to zero in on several elements that shape the global food trade.

Exploring the Roots of Agricultural Upset

On the macro level, food is not in short supply. According to a recent study, global food production can provide about 3000 calories per person per day, high above the essential 2000 calories needed (D'odorico et al. 2014). At the same time, however, food waste is an absolute failure at current levels. For instance, the United States Department of Agriculture (n.d.) reports that between 30 and 40 percent of the food supply is wasted each year, going into landfills instead of hungry families. This tragedy wastes labor and resources, and it also takes more costs to transport that waste to landfills. These are the events that we can study through measurements and statistics. The ones occurring at the level of individuals show that behind such misfortunes, real people suffer across the globe.

While some might think that the exploitation of farmworkers happens only in developing nations, such beliefs are incredibly mistaken. They are disrespected and harmed across the world. Workers in fields endure substantial harm and illness from pesticides (Payán-Rentería et al. 2012). These pesticides harm farmworkers' families (Abbasi-Jorjandi et al. 2020). Living conditions are often horrible. Due to legal loopholes designed for family farming operations, there is a history of seasonal US farmworkers having their living quarters substandard and dangerous where otherwise they would be prohibited (Arcury and Quandt 2011). Ironically and sadly, farmworkers must deal with food insecurity, too. Children as young as ten years old are legally permitted to work on farms under specified conditions in the US.

For instance, Thomas Arcury et al. (2021, 603) illustrate the extent of such conditions in the US:

> Although farm work is associated with high levels of pesticide exposure, and this pesticide exposure is associated with risk for immediate and long-term health and developmental problems, current federal labor regulations allow employers to hire children as young as 10 years of age for agricultural work in the US... . The number of children aged 10–17 years hired to work on farms each year is not known. The best available estimates indicate that between 30,000 and 79,325 hired child farmworkers were employed annually between 2005 and 2016.[3] Most of these child farmworkers are Latinx, either immigrants themselves or the children of immigrants. Agriculture is a hazardous industry, and children working in agriculture experience high rates of injury, illness, and death.

Though children are exposed to the conditions above, many more adults are, of course, and they must endure harm and put up with discrimination. Due to many social circumstances (e.g., political, economic), these workers cannot speak out against their environment. They cannot financially afford to protest or complain about labor conditions because they lack protection. For example, in the United States, many farmworkers remain on the margins of society, and the rural landscapes that accompany agricultural labor are apparent settings. These jobs come with minimal pay and often lack benefits, depriving them of essential medical, dental, and vision coverage. Yet, this kind of work takes a significant toll on workers' bodies and minds (Arcury and Quandt 2011).

While many laborers accept these jobs, arguably due to less desired socio-political-economic circumstances elsewhere, they maintain at least a semblance or degree of agency in most instances. Even though their labor conditions suggest otherwise, these workers also have dignity and human rights—and they deserve respect no matter what. Yet, while the circumstances above take place within the US, many people will be surprised to learn that actual slavery and forced-child labor practices are contemporary issues for agricultural work in select locations across the globe (Barclay 2013). Leigha Crout (2018, 1–2) states the conditions that make such outrages possible:

> Among the most influential participants in the modern global economy, perhaps no commodity is more highly valued than that of labor. This is

> exceptionally true for the industry of agriculture, the world's largest employer. At its current estimate, agriculture now employs over one-third of the international working population. However, in tandem with economic growth and the rising expansion of higher-income countries is the chronic shortage of farmworkers; as countries grow, domestic interest in the agricultural sector lapses into a steady decline. To solve this crisis of need, nation states have increasingly sought the engagement of migrant workers, who are hopeful for the opportunity to enter a more economically sound state to support their families abroad. However, this alternative has not elicited the desired effects; instead, it has facilitated the creation of a system crafted for the regulated exploitation of vulnerable groups.[4]

This passage illustrates the conditions associated with agribusiness that are arguably invisible and unknowable by most consumers across the globe. Understanding the dynamics behind the global economy takes advanced study and comprehension to grasp. It seems unlikely that World Geography courses cover these topics in primary schools.

Crout's review and assessment of the above issue paints with broad strokes. Still, it provides the necessary glimpse of food realities that global agribusiness inherently shields consumers from (a condition called "globalized opacity" that I cover in the next few chapters). The extent of harm that many people consider morally repugnant remains entirely removed from the final products on supermarket shelves. Moreover, the extent of such issues compounds such matter even further, suggesting that such harms are not exclusively outliers. Instead, such instances account for a significant proportion, even though any number of abuses is too many. For example, Crout (2018, 2–3) situates the nature and scope of the problem succinctly:

> Of the estimated 16.7 million migrant workers in the global agricultural sector, nearly 2 million are in situations of forced labor. For others, sexual violence, withholding wages, child labor, physical abuse, and dangerously unsafe environments are all well-documented daily occurrences. Rather than invest in much needed regulatory programs safeguarding migrant workers from abuse, governments tend to ignore these particular offenses. Few farmworkers receive the protection of trade unions, and an even smaller number complain of violations of labor laws. The small number of complaints can be attributed to two primary responses; fear of workplace retribution, or, in the case of irregular migrants, fear of detainment or deportation.[5]

This notion challenges the idea that such immoral practices disappeared from the planet. Still, researchers and activists have connected forced labor to the ingredients in popular consumer products that found their way to supermarkets. This reality entails that as people gathered around their family tables, they were indirectly connected to arguably the worst acts of humanity. Yet, it seems safe to assume that most of these eaters had no idea that they were participating in these offensive operations unknowingly.

Although the issues listed are not—by any means—a complete account of the hardships that field workers face, when considered together, they illustrate a line of work that shoulders much of the burden, which benefits business owners and consumers. The issues listed above show some of the worst practices that disrespect humanity. Yet, the reality that the problems above arise during the first stages of food production should provide a sobering view of what is to be expected when examining the totality of some food supply chains. After food is grown and harvested, it begins moving from one place to the next, with processing typically being the next stop. To gain insights into the nature of the issues that arise during these stops along the supply chain, the following section brings several such cases into view, highlighting areas of concern.

Between the Farm and the Table

Regarding food handling, one could hold that nothing will ever compare to Upton Sinclair's *The Jungle*'s impact on the public. It features horrific tales of the meatpacking industry, including humans becoming meat, rotten meat being processed, and child exploitation (Sinclair and Lee 2003). While this work revealed the grotesque nature of meat processing, it was mostly fiction. Still, it brought numerous dark issues to light, including but not limited to child labor, filthy work areas, processing of meat unfit for consumption, and horrible working conditions. In turn, the meatpacking industry met monumental reforms, improving the industry significantly.

Despite having a significant impact on meat processing in the US, other countries across the globe still face similar working conditions, such as children working long hours in unsafe conditions. These foods make their way across oceans, and consumers lack insights into these conditions. Such problems permeate the food industry today. While multinational food conglomerates have tried to eliminate connections to this abhorrent practice, food products tied to slavery have made their way onto supermarket shelves in stores such as Wal-Mart, one of the largest retailers in the world (Barclay 2013).

While the issues mentioned above indicate the kind of issues associated with some of the most troubling aspects of food processing, several other matters color food supply chains at this stage. Dangers come from accidents, chemical exposure, and dangerous equipment. In the US, people working in food processing remain subject to significant injuries and some deaths (Bureau of Labor Statistics 2021). While one could argue that removing all hazards from food processing is impractical, putting a higher demand on the industry than other sectors, the fact remains that food-processing occupations carry dangers that require attention. If we were to claim that they called for no examinations, then the previous ills likened to *The Jungle* would remain relevant. However, massive food production reforms that followed in the US and abroad illustrate that businesses can improve workers' safety—if regulations make them. These issues cut across cultures, governmental systems, religions, environments, and economic systems. Aside from having vastly different circumstances, the nature of the injustices and harms they contend with fall within the same range.

At the end of the food supply chain is the consumer. As mentioned earlier, they are likely unaware of the harms and dangers found along food supply chains, but they are often enduring harms stemming from food processing. For instance, while food safety issues remain a constant concern for agribusiness, the health aspects of food play a significant role in shaping the quality of life for people worldwide. Many of today's foods are unhealthy, and the availability of foods can influence how and what people eat, leading to adverse health impacts on the public (Guthman 2011). Still, individuals who want to eat healthily face numerous challenges to achieve desired outcomes.

For instance, learning about the intricacies of food ingredients and their relations to health outcomes is a daunting enterprise. While most food ingredients are healthy or at least benign, high fructose corn syrup (HFCS) is a controversial item found in many foods. While it is no doubt harmless in small quantities, it is linked to obesity and diseases such as diabetes. A surprising aspect of this ingredient is that research shows that, at least in the US, the federal government subsidizes it, making it a cheap option for businesses to increase their bottom lines (Fields 2004; Finkelstein and Zuckerman 2010; Franck et al. 2013; Epting 2017). Even worse, research shows that it harms the African-American population disproportionately (Saab et al. 2015). The point here is not to isolate HFCS as the "bogeyman." Yet, the ubiquitous nature of this ingredient shows how a simple sweetener added to food can insidiously find its way through food chains to our plates, lunch boxes, and drive-thru windows.

Along with these kinds of issues, there are also problems that, while not inherently part of food chains, intersect with them in ways that introduce additional aspects that remain external to topics such as food security. "Food deserts" are one such subject. In the US, due to how some cities are laid out, planned, or emerged, there are locations where healthy and affordable food is out of immediate reach for some populations that are already struggling in many cases. In turn, getting food to these residents remains challenging, requiring advanced efforts that go beyond food logistics. In turn, urban planning and municipal agents find themselves dabbling into food distribution schematics as part of their jobs, assuming they care in the first place.

One could argue that consumers are responsible for understanding what they are eating (or where they live in proximity to grocers). However, in the United States, the Centers for Disease Control and Prevention (2021) show that the educational requirements needed to create or correct healthy eating habits are significantly lacking. The total number of schools offering these educational measures has also declined. This point suggests that the type of knowledge required to eat in an optimally healthy manner remains out of reach for most consumers who do not have the time or desire to engage in advanced studies of the micro-level composition of today's foods. In turn, there is an argument here, holding that all humans involved in food chains have stakes in the quality of goods that can travel short, mid-range, or extremely long distances with numerous nonhuman actors such as multinational food companies behind the wheel. Yet, the views presented thus far merely paint in broad strokes to illustrate the extent of the problems associated with agribusiness.

The previous sections show how humans, from indigenous farmers to laborers, must deal with troubling conditions associated with agribusiness. Still, the focus is narrow, only paying attention to anthropocentric concerns. This point suggests that, even though humans cannot escape only having a human perspective, the nonhuman world deserves consideration. Nonhuman animals and ecosystems, for instance, have their own intrinsic value, and this condition warrants at least some degree of moral respect. This notion does not mean that we must have nonhuman interests ahead of humans, but it does entail bringing such ideas into our purview when dealing with those kinds of affairs. In turn, the following section investigates how agribusiness affects the nonhuman world in ways that significantly disregard its intrinsic value in inherently problematic ways.

Agribusiness and Nonhuman Animals and Ecosystems

While humans can fight against wrongdoing and bad outcomes that come from agribusiness, nonhuman nature cannot. This situation does not mean that wrongdoing is absent from such cases. Still, it suggests that we must examine such scenarios to determine what needs to change so that nonhumans can exist in a manner that shows respect, at least for its own sake, and because humankind requires nonhuman nature's instrumental value, which includes the mere ability to produce food. Some people might not care about nonhumans' intrinsic value, but not caring does not dismiss any moral obligations associated with this kind of value.

Yet, there are numerous kinds of situations showing how agribusiness harms and disrespects nonhuman life. The point here is not to put humankind beneath nature. We must produce food, obviously, and nonhuman systems require extreme manipulation for us to engage in such practices. Arguing that we should not grow food in select instances to appease nonhuman nature would be a tough sell for many people. Still, some methods could yield better outcomes. The goal, then, is to identify the sort of instances that would benefit from advanced thinking that can maximize the utility of nonhuman nature, which is the product of agribusiness.

Recent history is not in short supply of examples. For instance, perhaps the best-known case is factory farming. It has received numerous and extensive criticisms, exposing numerous kinds of harm to nonhuman animals (Foer 2009; Kemmerer 2015).[6] As mentioned earlier, pesticide runoff harms marine life in rivers, tributaries, and streams (Cooper 1993, Carpenter et al. 1998). The dead zone in the Gulf of Mexico is a phenomenon that challenges farmers, policymakers, and society (Greenhalgh and Faeth 2001). Deforestation is a growing concern in terms of raw land for production (Benhin 2006). Species displacement from appropriating land is another issue.

In general, biodiversity and its many benefits for ecosystems continue to dwindle, a worsening trend and some agribusiness harms or threatens endangered species (Conceição et al. 2022). Freshwater systems also feel the strain to irrigate crops, which has myriad impacts on nonhuman species in some cases that call for mitigation (Moore et al. 1996). Again, the point here is not to provide an exhaustive account of each instance wherein food production harms nonhuman species and ecosystems. Instead, it illustrates the kinds of issues that one can expect to encounter when examining agribusiness's relations with the nonhuman world.

At this point, it is worth mentioning that not all relationships or harmful impacts cause alarm. That is to say, while I want to argue that we should look at agricultural businesses with a keen eye for arguments for intrinsic value, such views require somewhat nuanced attention.[7] The idea is not that food production should be avoided in *all* locations where there is an ecological worry. Such a view could be catastrophic, causing starvation, economic turmoil, or any number of unforeseen harmful consequences. Instead, such situations will call for multidisciplinary assessments from a wide range of researchers in conversation to ensure that we are not overprotecting an ecosystem to the detriment of humankind, now or in the future.

This last point, distant generations of people, brings the notion of sustainable agriculture into view. Agribusiness plays a significant role in planetary climate change, which impacts people now, and scholars predict that its effects will worsen in the years and decades ahead (Sivakumar 2021). This point suggests that the totality of decisions and actions that shape food production and distribution will have far-reaching impacts that could help determine future people's quality of life. This significant challenge is essentially sustainable agriculture's internal tension, trying to meet today's needs without compromising distant people's ability to do the same.

This last point signals that sustainable agriculture could endure a prolonged or semi-permanent state of flux, considering that the effects of climate change will require significant study to determine how it will impact a changing planet. Such a period could also require a shift in agribusiness, perhaps moving some food-production practices back into urban centers, efforts already taking place at various scales in several cities (Ikerd 2017). In turn, metropolitan environments will have to choose between competing interests for limited land use. While water wars between farmers and cities are nothing new, the future could increase tensions between urban needs (Flörke et al. 2018).

For example, affordable housing issues plague numerous cities worldwide in highly complex ways (Daud et al. 2017). The places are already dealing with these kinds of competing interests. Adding another vital element, such as food production, could easily complicate these matters. This point is not meant to deter discussion about bringing food production into cities. However, it does suggest that doing so could face multiple challenges that will change depending on given cities' socio-political material characteristics. That is, what appears to be a solution for one location might not work so well in the next, meaning that measures aimed at developing solutions to such problems would be suggestive more than they would be universal.

This notion does not entail that we should not strive for limited measures. On the contrary, developing mitigatory and anticipatory measures that can help balance competing interests should remain firmly in view—optimizing such outcomes. Yet, chasing a universal, one-size-fits-all approach or device should be thought about with caution. The reason is that we do not serve ideas. We want them to work for us. Thinking about these two notions in tandem entails that if a policy or technology can deliver desired outcomes, pursuing them with a hopeful—yet evaluative—eye should hold steady.

The next chapter begins this process. It illustrates how specific theoretical devices can help us identify problematic issues that have and will continue to emerge within the complexity of the global food trade. In turn, this idea means that it starts by examining the ontological character of agribusiness. The thinking behind this approach is to show how we can better understand the problematic issues that come with a highly complex process that we must attend to keep the world functioning so that people can lead flourishing lives.

Conclusion

This chapter drew attention to the conditions of food production, starting with farming operations to get a sense of how agribusiness affects laborers the world over. It also focused on the dangers inherent to food processing. Exploring how the food trade harms people exhibits that such hazards do not stop once the products reach store shelves. They continue to endanger people while momentarily balancing on forks, sitting on spoons, and clenching between chopsticks. The reality that public health outcomes show that, despite having an abundance of food options, preservatives to keep food fresh, and low prices, today's food system leads countless people subject to ill effects directly linked to their proliferation (Guthman 2011).

Though humans suffer to varying degrees, the nonhuman world does as well. From deforestation to species displacement to global climate change, agribusiness significantly influences all such affairs. There is little doubt that it will continue to do so into the distant future, impacting several generations to come. Despite this grim outlook, developing measures to mitigate existing harm while doing our best to anticipate additional problems must hold steady as worthwhile goals that advance alongside food technologies. This notion suggests that the means that we employ must have incredible abilities to deliver better results for tomorrow.

Notes

1 There are numerous ways to document this point, each offering an informative perspective on the relationships between food production, distribution, and consumption and their role in global climate change.

2 It is worth mentioning that many of the issues covered in this chapter rest on the reality that numerous other historical atrocities remain hidden. In turn, we are not able to fully grasp the vast extent of injustice associated with food production and distribution. This point in no way suggests that such topics do not warrant investigation regarding how they connect with existing food networks. The work in this book addresses existing food chains, but this focus does not dismiss the significance that such issues have for the present. However, giving these issues the attention and focus that they deserve falls beyond this book's parameters. For more information, see Knobloch (1996).

3 United States Government Accountability Office (2019); Centers for Disease Control and Prevention (2019).

4 Due to aesthetic considerations, references for the sources that Crout (2018) uses are found in the original document listed in the References section of this chapter.

5 Again, due to aesthetic considerations, references for the sources that Crout (2018) uses are found in the original document listed in the References section of this chapter.

6 It is worth mentioning that factory farming in agribusiness shares ground with the animal rights' literature. That area of study gives ample attention to topics that require entirely separate conversations. In turn, I only mention them here broadly. The authors mentioned in the text cover them extensively. See Foer (2009) and Kemmerer (2015) for more information.

7 For a taxonomy of intrinsic value in an environmental-ethics context, see Hargrove (1992).

References

Abbasi-Jorjandi, M., Asadikaram, G., Abolhassani, M., Fallah, H., Abdollahdokht, D., Salimi, F., Faramarz, S., & Pournamdari, M. (2020). Pesticide exposure and related health problems among family members of farmworkers in southeast Iran: A case-control study. *Environmental Pollution*, 267, 115424.

Arcury, T.A., Chen, H., Arnold, T.J., Quandt, S.A., Anderson, K.A., Scott, R.P., Talton, J.W. & Daniel, S.S. (2021). Pesticide exposure among Latinx child farmworkers in North Carolina. *American Journal of Industrial Medicine*, 64(7): 602–619.

Arcury, T.A., Quandt, S.A., & Russell, G.B. (2002). Pesticide safety among farmworkers: Perceived risk and perceived control as factors reflecting environmental justice. *Environmental Health Perspectives*, 110(2), 233–240.

Arcury, T.A., & Quandt, S.A. (2011). Living and working safely: Challenges for migrant and seasonal farmworkers. *North Carolina Medical Journal*, 72(6), 466–470.

Barclay, E.(2013). Why slave labor still plagues the global food system. The Salt, June 20. https://www.npr.org/sections/thesalt/2013/06/19/193548623/why-slave-labor-still-plagues-the-global-food-system(accessedMay 29, 2021).

Benhin, J.K. (2006). Agriculture and deforestation in the tropics: A critical theoretical and empirical review. *AMBIO: A Journal of the Human Environment*, 35(1), 9–16.

Bureau of Labor Statistics(2021). Employer-reported workplace injuries and illnesses – 2020. https://www.bls.gov/news.release/pdf/osh.pdf (accessed April 20, 2022).

Carpenter, S.R., Caraco, N.F., Correll, D.L., Howarth, R.W., Sharpley, A.N., & Smith, V.H. (1998). Nonpoint pollution of surface waters with phosphorus and nitrogen. *Ecological Applications*, 8(3), 559–568.

Centers for Disease Control and Prevention (2019). Childhood agricultural injury prevention initiative. https://www.cdc.gov/niosh/topics/ childag/cais/techinfo.html#Datasource (accessed April 20, 2022).

Centers for Disease Control and Prevention (2021). Nutrition education in US schools. https://www.cdc.gov/healthyschools/nutrition/school_nutrition_education.htm (accessed February 2, 2022).

Conceição, E.O., Garcia, J.M., Alves, G.H.Z., Delanira-Santos, D., de Fátima Corbetta, D., Betiol, T.C.C., ... & do Couto, E.V. (2022). The impact of downsizing protected areas: How a misguided policy may enhance landscape fragmentation and biodiversity loss. *Land Use Policy*, 112, 105835.

Cooper, C.M. (1993). Biological effects of agriculturally derived surface water pollutants on aquatic systems—a review. *Journal of Environmental Quality*, 22(3), 402–408.

Crout, L. (2018). Farmworkers worldwide: The invisible citizens. Master's Thesis, Cornell University.

Daud, N.M., Nor, N.M., Ali, U.N., Yusof, M.A., & Munikanan, V. (2017). Affordable housing system: A review on issue of housing affordability. *The Social Sciences*, *12*(7), 1281–1287.

D'odorico, P., Carr, J.A., Laio, F., Ridolfi, L., & Vandoni, S. (2014). Feeding humanity through global food trade. *Earth's Future*, 2(9), 458–469.

Epting, S. (2017). Introduction to food justice and the built environment. In *Food justice in US and global contexts* (pp. 89–93). Cham: Springer.

Finkelstein, E.A., and Zuckerman, L. (2010). *The fattening of America: How the economy makes us fat, if it matters, and what to do about it.* Hoboken, NJ: Wiley.

Fields, S. (2004). The fat of the land: Do agricultural subsidies foster poor health? *Environmental Health Perspectives*, 112(14): A820.

Flörke, M., Schneider, C., & McDonald, R.I. (2018). Water competition between cities and agriculture driven by climate change and urban growth. *Nature Sustainability*, 1(1), 51–58.

Franck, C., S.M. Grandi and M.J. Eisenberg. (2013). Agricultural subsidies and the American obesity epidemic. *American Journal of Preventive Medicine*, 45(3): 327–333.

Foer, J.S. (2009). *Eating Animals*. New York: Little, Brown and Company.

Gilson, E., & Kenehan, S. (2018). Introduction. In E. Gilson & S. Kenehan (Eds.), *Food, environment, and climate change: justice at the intersections*. London: Rowman & Littlefield International.

Greenhalgh, S., & Faeth, P. (2001). A potential integrated water quality strategy for the Mississippi River Basin and the Gulf of Mexico. *The Scientific World Journal*, 1, 976–983.

Guthman, J. 2011. *Weighing in: Obesity, food justice, and the limits of capitalism.* Oakland, CA: University of California Press.

Hargrove, E.C. (1992). Weak anthropocentric intrinsic value. *The Monist*, 75(2), 183–207.

Ikerd, J. (2017). THE ECONOMIC PAMPHLETEER: The urban agriculture revival. *Journal of Agriculture, Food Systems, and Community Development*, 7(3), 13–16.

Kemmerer, L. (2015). *Eating earth: Environmental ethics and dietary choice.* Oxford: Oxford University Press.

Knobloch, F. (1996). *The culture of wilderness: Agriculture as colonization in the American West.* Chapel Hill, NC: University of North Carolina Press.

La Via Campesina. The global voice of peasants! https://viacampesina.org/en/wp-content/uploads/sites/2/2021/12/LVC-EN-Brochure-2021-03F.pdf.

London, L. (2003). Human rights, environmental justice, and the health of farm workers in South Africa. *International Journal of Occupational and Environmental Health*, 9(1), 59–68.

Metcalfe, R. (2019). *Food routes: Growing bananas in Iceland and other tales from the logistics of eating.* Cambridge, MA: MIT Press.

Moore, M.R., Mulville, A., & Weinberg, M. (1996). Water allocation in the American West: Endangered fish versus irrigated agriculture. *Natural Resources Journal* 36(20): 319–357.

Payán-Rentería, R., Garibay-Chavez, G., Rangel-Ascencio, R., Preciado-Martinez, V., Munoz- Islas, L., Beltrán-Miranda, C., Mena-Munguía, S., Jave-Suárez, L., Feria-Velasco, A., & De Celis, R. (2012). Effect of chronic pesticide exposure in farm workers of a Mexico community. *Archives of Environmental & Occupational Health*, 67(1): 22–30.

Saab, K., Kendrick, J., Yracheta, J., Lanaspa, M., Pollard, M., & Johnson, R. (2015). New insights on the risk for cardiovascular disease in African Americans: The role of added sugars. *Journal of the American Society of Nephrology*, 26: 247–257.

Schwartz, N.A., von Glascoe, C.A., Torres, V., Ramos, L., & Soria-Delgado, C. (2015). "Where they (live, work and) spray": Pesticide exposure, childhood asthma and environmental justice among Mexican-American farmworkers. *Health & Place*, 32, 83–92.

Sinclair, U., & Lee, E. (2003). *The Jungle: The Uncensored Original Edition.* Tucson, AZ: See Sharp Press.

Sivakumar, M. (2021). Climate change, agriculture adaptation, and sustainability. In *Climate resilience and environmental sustainability approaches* (pp. 87–109). Singapore: Springer.

United States Government Accountability Office (2019). Working children: Federal injury data and compliance strategies could be strengthened. https://www.gao.gov/assets/700/695209.pdf (accessed April 20, 2022).

3 Food Chains and Applied Mereology

Introduction

The last chapter illustrated the issues associated with a food chain's links, or its "parts," affecting several stakeholders. Despite such problems, we need agribusiness, but we also need to rid it of the undesired outcomes previously discussed. Although a worldwide network connects multiple parts of conventional food chains to facilitate international commerce, this process produces "globalized opacity." This term means that there is an overabundance of components and often long distances between many of them. However, considering the reality that having numerous parts is a complicated matter and that these parts are often sold globally, traveling an incredibly long distance is not an essential condition per se when considered within an entire food chain.

There will always be outliers. Still, the topic here is about the food chains found worldwide. The sheer number of them means that there will be ample variety. Yet, considering the consolidated nature of the global food trade, a recognizable pattern emerges, one wherein we can see a vast number of exchanges occurring to yield international sales. Due to the worldwide market and its sales, we see individual micro ingredients as parts become part of larger, macro ingredients used to produce food productions ready for consumption (Metcalfe 2019). This process occurs across the world innumerable times a day to keep store shelves stocked, restaurants cooking, and meals being served on planes, in hospitals, and delivered to hungry, homebound seniors.

Considering this globalized context, a food chain is an entity where the links are spread across vast distances. Due to this situation, most people cannot fully see and know about the composition of a conventional food chain. Somewhat paradoxically, the parts that make it feasible for these components to be noticeably disconnected, namely transportation and logistics—are also pieces of the farm-to-table puzzle.

Moreover, these parts are indispensable for today's food chains. Yet, when it comes to traditional views of farming, these elements remain mostly unknown. This condition makes it exceedingly difficult to know about many

DOI: 10.4324/9781003255505-3

issues connected to conventional food supply chains, which is paramount when considering several wrongs associated with them. Attending to such matters could help us see them more clearly, which would bolster efforts to understand the part-to-part and part-to-whole relations within a food chain.

Gaining a clear view of the nature of this issue could improve how food-justice researchers understand the intricacy involved in these affairs, which could assist struggles to alleviate harm. One way to support their efforts is to employ an "applied-mereological" approach, which helps us examine the relations of a food chain's parts, locating the parts we cannot readily know about. This kind of investigation provides a new measure that can help us understand how conventional food chains are problematic. In turn, the purpose of this chapter is to move in that direction. The goal is to identify part-to-part relations that favor ethical outcomes. A significant step to advancing that goal is to show how the problem of globalized opacity plays a role in such affairs.

To make this case, I begin by reviewing how we can employ the basic structure or idea behind mereology to help us understand the composition of conventional food chains. Specifically, I argue that we can better understand that food chains adhere to "unrestricted mereological composition," meaning that any part, concrete or abstract, can qualify as a part of a food chain.[1] In turn, creating this positioning provides a way to employ this thinking to narrow the focus toward mitigatory efforts, which we explore in the next chapter.[2]

The Benefit of "Applied Mereology" for Understanding Conventional Food Chains

In a basic sense of the theory and its study, mereology deals with the interplay of parts, along with how parts relate to wholes (Varzi 2015). Within the traditional literature, philosophers who engage in this research address highly abstract relations of parts (Hovda 2009). While the nature of these undertakings might not interest scholars outside of this field, this view is shortsighted. That is, one benefit of employing this approach is that it helps us understand how objects that are composed of smaller and frequently overlapping parts fit together, forming a cohesive unit (Paul 2002). Due to the composition of food chains that involve numerous parts, interdisciplinary researchers can benefit from mereology, and the study of the conventional food chain supports this claim.[3]

From the outset, the necessary and sufficient conditions for thinking about a food chain's parts in such a manner requires that for parts, they must be parts in and of themselves, and they must also be a part of the food chain. Complicating matters a bit further, some parts are parts of themselves, while also being parts of greater wholes, which are also parts of food chains. Consider this brief example: land, farm equipment, and people are parts themselves, and they help compose a farm, and that farm is part of a food chain.

For instance, highly detailed illustrated maps show how the numerous parts of a conventional food chain fit together and interact with each other, revealing the dynamic character of how food supplies crisscross the globe.[4] Although these maps appear stable, Jo Goossens, the person who created one of the most intricate maps to date that lays out the vital exchanges within global agribusiness, points out that micro-level aspects of the food trade remain in constant flux (Cereals & Grains Association 2020). This notion suggests that literal parts of conventional food chains are continuously adapting to abstract parts such as market forces and labor, along with other literal parts such as new technologies, advances in logistics, and the nature of business (Regmi and Gehlhar 2005).

These chains are historically situated, meaning that entities such as the multinational food conglomerates that control conventional food chains are entrenched deeply in the processes and exchanges that produce most of the world's food supply (Howard 2016). While this notion might sound trivial, it should be significant, considering that people can shape food chains through supporting legislative efforts (Mars and Ball 2016). In turn, going from the farm to the table is not a simple matter, and accounting for and explaining the process is not much easier. Further, it should not come as a surprise that most consumers in countries such as the United States have only a vague idea of the direct origins of their food (Blatt 2011).

Consider, for instance, that the outline of a food chain is quite encompassing, consisting of numerous concrete parts, often crossing national and cultural boundaries (Metcalfe 2019). At present, ten companies control over half of the food supply in the United States, extending to about 15 percent of the global market (Lyson and Raymer 2000; Stuckler and Nestle 2012). Considering that the worldwide food exchange accounts for 10 percent of the world's total economy, estimates show that the global food supply is worth at least 8 trillion dollars annually (Van Nieuwkoop 2019).

Although, as indicated above, the parts change due to several considerations, and the control of the parts remains competitive, extending into numerous areas such as economics and international political affairs. Due to such conditions, when considering food chains and the kind of parts that they can have, this reason shows why it is imperative to think about food chains as having an unrestricted mereological composition.[5] In turn, the way a food chain is presented here is one wherein some parts are strict while others are metaphoric, yielding an idealization with different concrete instantiations.[6]

To illustrate the ramifications of conventional food chains with these notions in mind, we must consider the range of their inventory and the scope of their impact as elements that require highlighting due to the array of effects they produce in concert. The point worth underscoring is that any change in a food chain's parts must bring financial realities into view. While it is evident that such alterations have epistemic values, one could hold that they also have a metaphysical correlate. If such changes stem from financial realities, then as abstract parts, finances can dictate a principle of composition for a food chain.

Here is an example. If a nation were to accept a loan from an international bank on the condition that it was to produce tea instead of soybeans, then the abstract part of financial pressure would alter the composition of the food chain by forcing concrete parts such as seeds and the necessary agricultural equipment to change. In such instances, parts are "stuck" together due to economic considerations. In turn, differing economic reasons can generate different principles of composition. This notion suggests that the ontology of such chains requires that researchers who examine their parts pay attention to their backstories (Howard 2016). The reason to emphasize this idea is that it is unreasonable to weigh the relationships between parts without bringing these vital dimensions into view, especially considering that numerous components require investigation.

This area would benefit from thinking about food chains regarding unrestricted mereological composition. For instance, we can break their inventory down into several categories of non-overlapping or disjoint parts, meaning the part is not counted more than once when accounting for its place within the whole food chain (Varzi 2000; Varzi 2014; Lando 2017). These parts include but are not limited to concrete parts such as land, water, natural resources for production, petroleum resources, and chemical fertilizers. Yet, they also include abstract parts such as labor forces, federal, state, and municipal regulations, communication networks to facilitate the logistics and distribution, which have literal parts such as transportation systems and storage facilities, marketplaces, and restaurants (Goodman et al. 2012; Morley and Marsden 2014; Pitt and Jones 2016; Ruben et al. 2019).

Considered together, these parts (along with several other parts if that is the case) can compose a greater part, which then becomes part of the larger whole, such as a food chain. For example, parts such as a farmer, her land, the water, and various agricultural products can compose a farm. This farm can be a part of a food chain. Depending on the farm's character (e.g., small, organic, or large family farm), it can have significant real-world impacts. By putting these aspects into real numbers, we see the requirements and effects of conventional food chains. For instance, Eva Gladek et al. (2017, 4) hold:

> Agriculture now occupies roughly half of the plant-habitable surface of the planet, uses 69% of extracted fresh water and, together with the rest of the food system, is responsible for 25–30% of greenhouse gas emissions. The expansion of industrial fishing fleets and a higher demand for seafood globally have led to the collapse or total exploitation of over 90% of the world's marine fisheries. A growing demand for land-based animal products is the primary driver of tropical deforestation. Through its direct and intermediate impacts, the food system is the largest contributor to the depletion of biodiversity.

While this passage exhibits the scope of conventional food chains' impacts on the planet, they also reveal the effects that some people might not consider, such as how food chains affect marine ecology. One could argue that the view of the oceans often remains secondary when stacked against land-based concerns, bearing in mind that most people do not have experiences with those environments. For instance, Clark Wolf (2003) argues that because most people do not typically engage with oceans, they do not have the necessary perspective to grasp their complex character. Yet, when examining the numbers above, the accumulating effects on aquatic chains remind us that considering humankind's involvement with food chains remains paramount.[7] Due to this condition, we cannot dismiss the reality that such environs are significant macro-parts of the food chain, including numerous smaller parts that compose the fishing industry.

This notion reveals that, in addition to how specific parts of the food chain affect the non-human world, they also play a dominant role in the socioeconomic parts of food chains. For instance, continuing with the insights from Eva Gladek et al. (2017, 4) above, such considerations contribute to a panorama of conventional food chains, exhibiting the significance of such dimensions:

> The agri-food sector is the world's largest economic sector and is therefore deeply entwined with poverty. Half the global workforce is employed in agriculture. A majority of the world's poorest people are subsistence farmers and fishermen. Small farmers and fishers around the world are caught in cycles of poverty, without access to education, employment, economic and social infrastructure, and political representation. Many do not receive adequate compensation, work in unacceptable conditions, or do not have access to sufficient, affordable, or proper-quality food. Poverty is the largest threat to producers of food globally and the largest driver of food insecurity.

The passage above shows that while we can separate concerns about food chains from considerations for other social systems, they require additional study due to the numerous effects that they can help produce. Philosophers of food and interdisciplinary scholars have documented these outcomes, analyzing the many ways that conventional food chains affect farmers and indigenous people on almost every continent by focusing heavily on the concept of food sovereignty (Jarosz 2014: Menser 2014; Navin 2014; Werkheiser and Noll 2014; Grey and Patel 2015; Epting 2018). The pattern of thought behind applied mereology can help us understand and manage issues of significance, such as food sovereignty. The section below illustrates this point. It shows how the complex nature of these problems could benefit from breaking food-chain issues down. In turn, the process makes it easier to identify the relationships of parts that stand in the way of food sovereignty and help find or create exchanges that support it.

Applied Mereology and Food Sovereignty

Viewing the food sovereignty literature as a compilation, the theme that emerges shows that numerous people in the agricultural industry seek to gain control over the parts that impact their lives. This notion pushes against the multinational food conglomerates, as mentioned earlier. In turn, if they were able to control the parts, then they would gain more control over the sum, a notion that appears to be consistent with the stated aims of some of the people who champion food sovereignty.

A recent embodiment of these sentiments has emerged from the labor force that works in food production, as mentioned in Chapter 2, La Via Campesina. Recall that this organization of farmers spans the globe, consisting of almost two hundred smaller organizations from over eighty countries in Asia, Africa, the Americas, and Europe (La Via Campesina 2007). Their concerted efforts promote smaller family farming practices toward agricultural sustainability and justice. This group employs the term food sovereignty as a central tenet to their approach to address how groups express autonomy over their involvement with all aspects of the food trade. They use it in a manner that brings numerous related concerns into view, including but not limited to social justice, safety, control, and human rights.

One can argue that they illustrate that several concerns are indirectly connected to conventional food chains within their conception of food sovereignty. Yet, many of these issues are not directly linked to the food industry, such as concern for future people (La Via Campesina 2007). The scope of this conception challenges the totality of the effects that conventional food chains help produce. In turn, La Via Campesina positions itself against the status quo, advocating for a means of agricultural production that can remedy the ill effects of globalized agriculture. For example, during the world's first conference on food sovereignty, they formulated the Declaration of Nyéléni (La Via Campesina 2007). It is a comprehensive statement that provides a panorama of the kinds of issues that we find in conventional food chains, along with several indirect concerns that also require attention:

> Food sovereignty is the right of peoples to healthy and culturally appropriate food produced through ecologically sound and sustainable methods, and their right to define their own food and agriculture systems. It puts the aspirations and needs of those who produce, distribute and consume food at the heart of food systems and policies rather than the demands of markets and corporations. It defends the interests and inclusion of the next generation. It offers a strategy to resist and dismantle the current corporate trade and food regime, and directions for food, farming, pastoral and fisheries systems determined by local producers and users. Food sovereignty prioritises local and national economies and markets and empowers peasant and family farmer-

> driven agriculture, artisanal-fishing, pastoralist-led grazing, and food production, distribution and consumption based on environmental, social and economic sustainability.

The image that brings the above dimensions into view is daunting in the best case and overwhelming in the worst case. They mention several points of concern, bringing current issues such as production and distribution into view. They also extend into future-oriented matters that involve future generations within a sustainability matrix. If we keep in mind that such areas involve numerous parts that serve as "actors," it is challenging to know the limits as to what counts as a part. One could argue that the description above reveals how the condition of globalized opacity that comes from the composition of conventional food chains can manifest through the felt impacts on the people who have been treated as only "parts," in the same manner as if they were another piece of farm equipment, a mere means to an end.[8] It could be the case that the declaration above exhibits that food workers do not feel respected in several ways. Bearing this point in mind, one could hold that they are responding accordingly, striving for measures that would rearrange or exchange parts so that they could control them, expressed as a call for food sovereignty. Considering that they would gain control over more parts, then this power would show that they are not merely parts such as farming equipment, which of course, cannot control other parts.[9]

Although the declaration above draws attention to injustices that arise due to "Big Food" having power over the parts, such situations remain challenging to identify due to the problem of globalized opacity. People are unaware that the parts of conventional food chains are arranged and managed in a fashion that makes it exceedingly difficult to know. For instance, it is common for people in countries such as the United States to lack basic knowledge about food production (Blatt 2011). If they lack this essential knowledge, then it seems plausible to hold that they also do not know about injustices that emerge from it, a situation that feeds globalized opacity.

One could easily object to the condition above. One could hold that there is an overabundance of parts, and farmers and food workers on the frontlines experience problems, but there is no way to prove that such issues are a result of the existence of the former. While this concern is formidable, we should have some reservations about it. For instance, the primary claim in this chapter holds that a mereologically informed approach aims to improve the situation so that we can see all of the parts to analyze them adequately. If there is no connection, as the objection above maintains, employing this approach would support that claim. If the opposite view emerges, we gain a method for developing a solution to such problems.

In addition to this point, the idea of "disparate impacts" holds that, for dealing with issues that concern historically marginalized groups, they are victims of systematic discrimination. This notion entails that topics such as intentionality do not matter for these fairs.[10] This condition shows that we need not find a direct association. The outcome wherein they are subject to harmful treatment is sufficient. The task then becomes to identify precisely how the chain creates harm and wrongdoing. In the case of conventional food chains and applied mereology, that endeavor is the motivation that underpins this exploration, to deliver a way to identify how the relations between food parts and the whole food chain yields rotten fruits.

Without tending to the epistemological conditions for knowing about food injustices, one could argue that the projected reality for the future of food justice remains grim, offering little hope for creating a meaningful alternative that can remedy the situation described earlier. This notion implies that further identifying the interplay between parts will require advanced study. Such an undertaking aims to determine appropriate actions to rearrange them. Hopefully, doing so will lead to outcomes consistent with the people who are arguably treated as merely and only as "parts" by the multinational food conglomerate. They own and or control the majority of the parts that belong to conventional food chains, and we cannot ignore this reality.

Within various geopolitical regions, along with traditionally grown foods, the scope of such a call would require specific and or hyperspecialized concrete and abstract parts, including but not limited to safety regulations, farm equipment, and distribution schemes that could redistribute parts ethically. In the cases of both the broad view and the site-specific instances mentioned above, the numerous parts have a role in the unjust outcomes that scholars criticize. The question here is not to ask about finding the link between "parts A–Z" and "outcome X." The idea is that because numerous parts remain in play, determining how different parts relate *is* possible, but the daunting nature of such a task is a significant deterrent. In turn, the condition of having numerous parts facilitates globalized opacity.

In the next chapter, I examine the roots and status of the situation, followed by an investigation into alternate models that inherently avoid globalized opacity. While attending to the latter matters is a comprehensive enterprise, properly situating it serves as the stage that makes it possible for us to see how it could play out concretely. Although much more work needs to be done on this front, mereologically and practically, addressing such dimensions can better position that problem so that we can better grasp several of its troubling intricacies.

Conclusion

Having provided examples of the issues found throughout food chains, this chapter worked toward the required efforts to mitigate harms and chart a beneficial way forward. Although food chains are highly complex, breaking them down to see them in their part-to-part relations is the first step to conceptualizing them in a manner geared towards improvements. One such direction is dealing with the condition of globalized opacity, a kind of situation that results from having numerous food-chain parts cross vast distances many times over. Then, a case was made, illustrating how the thinking behind applied mereology could be advantageous because it simplifies understanding the process.

The main benefit of employing this approach is how it frames issues as part-to-part relations and how parts relate to their larger parts. This case concerns how the parts of a food chain, the "links," work together to form a larger unit, the food chain. One primary element that came into view was it could bolster efforts to secure food sovereignty for the groups that desire it. After understanding this term and its significance for creating better food-related outcomes, the work here is positioned to deal with a range of issues that intersect with globalized opacity.

Notes

1 To gain an in-depth look at some of the issues pertaining to the thesis behind unrestricted mereological composition that are beyond the scope of this chapter, see McCarthy (2015).
2 While there are other issues with unrestricted mereological composition, I am only applying it to the context of food systems in this chapter.
3 Frederique de Vignemont et al. (2005) exhibit how we can use the mereology to account for the ways that we experience parts of our bodies to how we experience the body itself. Peter Simons (2013) examines mereology of engineering and AI.
4 For a highly detailed food map of the global food chain, see Jo Goossens and the Global Food Chains Map, available online: https://www.cerealsgrains.org/publications/cfw/2019/jan-feb/Pages/CFW-64-1-0010.aspx.
5 It is worth mentioning that making such identity claims could be controversial if one holds that such an identity is impossible. However, engaging in this discussion is beyond the scope of this chapter. For more information, see McCarthy (2015).
6 Some philosophers, such as Giorgio Lando, maintain that is incredibly challenging at times to differentiate literal and metaphorical parts. Due to this condition, the claims in this chapter avoid engaging in such detailed discussion that, while relevant, fall outside of the scope of this chapter. For more information, see Lando (2017).
7 This claim does not imply the oceans are merely food chains. However, the intention is to draw attention to the idea that people could view them in such a manner.
8 This point raises numerous ethical issues that are beyond the aim of this chapter. However, for more information, see Kant and Schneewind (2002).
9 This point introduces the idea that changing or rearranging parts could produce a new food system or help the old food system become more just. Although this point is extremely relevant, giving it the attention that it deserves falls outside of this book's limits.
10 The United States Supreme Court introduced the term "disparate impacts" in *Griggs v. Duke Power Co.*, 401 U.S. 424, 91 S.Ct. 849, 28 L.Ed.2d 158 (1971).

References

Blatt, H. (2011). *America's food: What you don't know about what you eat.* Cambridge, MA: MIT Press.

Cereals & Grains Association. (2020). Spotlight on Jo Goossens and the global food chains map. *Cereal Foods World*, 64(1): DOI: https://doi.org/10.1094/CFW-64-1-0010.

de Vignemont, F., Tsakiris, M., & Haggard, P. (2005). Body mereology. In G. Knoblich, I. Thornton, M. Grosjean, & M. Shiffrar (Eds.), *Human body perception from the inside out* (pp. 147–170). New York: Oxford University Press.

Epting, S. (2018). Advancing food sovereignty through interrogating the question: What is food sovereignty? *Journal of Agricultural and Environmental Ethics*, 31 (5), 593–604.

Epting, S. (2021). Urban infrastructure and the problem of moral praise. *Techné: Research in Philosophy and Technology*, 25(1), 112–129.

Gamboa, G., Kovacic, Z., Di Masso, M., Mingorría, S., Gomiero, T., Rivera-Ferré, M., & Giampietro, M. (2016). The complexity of food chains: Defining relevant attributes and indicators for the evaluation of food supply chains in Spain. *Sustainability*, 8 (6), 515.

Gladek, E., Fraser, M., Roemers, G., Sabag Munoz, O., Hirsch, P., & Kennedy, E. (2017). The global food chain: An analysis, *Metabolic*. Available online: https://www.metabolic.nl/publications/global-food-chain-an-analysis-pdf/.

Goodman, D., DuPuis, E.M., & Goodman, M. (2012). *Alternative food networks: Knowledge, practice, and politics*. Routledge: Abingdon.

Grey, S., & Patel, R. (2015). Food sovereignty as decolonization: Some contributions from indigenous movements to food chain and development politics. *Agriculture and Human Values*, 32, 431–444.

Hovda, P. (2009). What is classical mereology? *Journal of Philosophical Logic*, 38 (1), 55–82.

Howard, P. (2016). *Concentration and power in the food chain: Who controls what we eat?* (Vol. 3). London: Bloomsbury Publishing.

Jarosz, L. (2014). Comparing food security and food sovereignty discourses. *Dialogues in Human Geography*, 4(2), 168–181.

Kant, I., & Schneewind, J. (2002). *Groundwork for the Metaphysics of Morals*. Binghamton, NY: Yale University Press.

La Via Campesina (2007). Declaration of Nye 'le' ni, Se' lingue', Mali. February 27Available online: https://nyeleni.org/spip.php?article290.

Lando, G. (2017). *Mereology: A philosophical introduction*. London: Bloomsbury Publishing.

Lyson, T., & Raymer, A.L. (2000). Stalking the wily multinational: Power and control in the U.S. food chain. *Agriculture and Human Values*, 17(2), 199–208.

Mars, M., & Ball, A. (2016). Ways of knowing, sharing, and translating agricultural knowledge and perspectives: Alternative epistemologies across non-formal and informal settings. *Journal of Agricultural Education*, 57(1), 56–72.

McCarthy, T. (2015). A note on unrestricted composition. *Thought: A Journal of Philosophy*, 4(3), 202–211.

Menser, M. (2014). The territory of self-determination: Social reproduction, agroecology, and the role of the state. In P. Andree, J. Ayres, M. Bosia, & M. Massicotte (Eds.), *Globalization and food sovereignty: Global and local change in new politics of food*. Toronto: University of Toronto Press.

Metcalfe, R. (2019). *Food routes: Growing bananas in Iceland and other tales from the logistics of eating*. Cambridge, MA: MIT Press.

Morley, A., & Marsden, T. (2014). Current food questions and their scholarly challenges: Creating and framing a sustainable food paradigm (pp. 17–45). In *Sustainable Food Chains*. Abingdon: Routledge.

Navin, M. (2014). Local food and international ethics. *Journal of Agricultural and Environmental Ethics*, 27, 349–368.

Paul, L. (2002). Logical parts. *Noûs*, 36(4), 578–596.

Pitt, H., & Jones, M. (2016) Scaling up and out as a pathway for food chain transitions. *Sustainability* 8, 1–16.

Regmi, A., & Gehlhar, M. (2005). *New directions in global food markets.* Washington, DC: U.S. Department of Agriculture.

Ruben, R., Verhagen, J., & Plaisier, C. (2019). The challenge of food chains research: What difference does it make? *Sustainability*, 11(1), 1–14.

Simons, P. (2013). Varieties of parthood: Ontology learns from engineering. In *Philosophy and engineering: Reflections on practice, principles and process* (pp. 151–163). Dordrecht: Springer.

Stuckler, D., & Nestle, M. (2012). Big food, food chains, and global health. *PLoS medicine*, 9(6), 1-4.

Van Nieuwkoop, M. (2019). Do the costs of the global food chain outweigh its monetary value? *World Bank Blogs*. Available online: https://blogs.worldbank.org/voices/do-costs- global-food-chain-outweigh-its-monetary- value.

Varzi, A. (2000). Mereological commitments. *dialectica*, 54(4), 283–305.

Varzi, A. (2014). Counting and countenancing. In *Composition as Identity* (pp. 47–69). Oxford: Oxford University Press.

Varzi, A. (2015). Mereology. In Edward N. Zalta (Ed.), *The Stanford encyclopedia of philosophy*. Available online: http://plato.stanford.edu/archives/spr2015/entries/mereology/.

Werkheiser, I. and Noll, S. (2014). From food justice to a tool of the status quo: Three sub-movements within local food. *Journal of Agricultural and Environmental Ethics*, 27, 201–210.

Wolf, C. (2003). Environmental ethics and marine ecosystem: From a "land ethic" to a "sea ethic." In *Values at Sea* (pp. 19–32), Athens, GA: University of Georgia Press.

4 Globalized Opacity

Introduction

After establishing how we can apply the thinking behind mereology, this chapter examines the broad scope of parts, zeroing in on the problem of globalized opacity associated with such arrangements. Through employing this term as a theoretical device, we gain a novel technique for conceptualizing a specific condition that we find with these chains. With an understanding of the magnitude of their effects, I explore some of the reasons why activists, philosophers, and food scholars challenge and want to change the status quo, advocating for alternatives that could have fewer parts and or less globalized opacity. The point that requires substantial underscoring is not that an abundance of parts or their opaque nature is inherently problematic. However, such conditions could, in many cases, facilitate harmful outcomes. With this idea in mind, the task is to examine part-to-part relations to determine if and how both are troublesome or require further investigation. This chapter aims to flesh out these notions, exposing some vital areas that would benefit from additional study.

The Conditions Surrounding Globalized Opacity

The globalized opacity of conventional food chains is not a condition that simply comes from only having multifaceted operations. Such processes arise due to an overabundance of parts from scattered locations, primarily connected through advanced logistics and transport. Framing this issue in such a manner reveals why using a mereologically informed approach is beneficial for gaining a clearer understanding of the relations that pertain to how the parts of a food chain fit together to form the composite object, "food chain." It also shows why developing mitigatory steps to food issues remains challenging. It focuses on the relations between parts and how the parts of a food chain impact the food chain itself.

That is, as food scholars such as Metcalfe (2019) exhibit, the conventional food industry is inherently global, suggesting that the very nature of food requires an encompassing lens to see its expansive character. It is not just

DOI: 10.4324/9781003255505-4

that our foods are made from ingredients from far-away places. The issue is incredibly more extensive than we can grasp initially. For instance, a simple ingredient can come from numerous other ingredients that crisscrossed the globe. This point suggests that what we consider a primary ingredient was already global before it made its way into the marketplace, which is, of course, global. In turn, the ways that we think about today's food are complicated from the outset, yet we must develop conceptual devices that are highly efficient without producing superfluous information.

For example, when talking about the globalized opacity of conventional food chains, each word is itself a smaller "part" of a more extensive theoretical device—the "globalized opacity of conventional food chains."[1] Each word in this phrase, as a conceptual device, plays a role in how we understand the reality of the situation, which are the conditions that come from having an overabundance of concrete and abstract parts. The precise nature of the mechanics of the phrase, as a theoretical device, reveals the pattern that we are ultimately addressing. If we have too few words (or theoretical parts that make up the larger theoretical device), we would not properly emphasize the exact area of study.

For instance, if we were to say that we are only dealing with "food chain complexity," then we would only be highlighting the notion that the food chain has a relationship status that reads: "it is complicated." Yet, such a scenario does not show that it is convoluted because there are simply too many parts involved in a multifaceted process. Being highly sophisticated is a separate issue from having an overabundance of parts. Although there are similarities between "food chain complexity" and "globalized opacity of conventional food chains"—theoretical devices that help us understand the issues they represent—the primary difference is that each tool does different tasks. The former focuses heavily on the processes that affect the interrelations between parts. The latter pays much more attention to parts in terms of how they affect others as components of the food chain and the food chain itself.

Consider, for instance, that Gamboa et al. (2016, 2–3), in their research on the complexity of food systems, make this notion apparent:

> Owing to the many domains involved and the different scales on which different processes take place (from households to the global market), food systems are inherently highly complex systems: That is, their relevant aspects cannot be captured from a single perspective, and therefore different stakeholders may have different perceptions of what a food system is and how it performs.

When examining the passage above, we see that the operational relations of food chains are at the forefront of their approach, revealing the prominent position processes have in food chains. Namely, within the short description above, two instances show action. The first exhibits that the *processes* of scale help define a food chain's complexity. The second notion concerns not relying on a single perspective of a food chain's definition and *performance*. While the passage above does not represent all accounts that address food-chain complexity, one could argue that such approaches are typical.

This point does not discount the notion that parts interact in specific and numerous ways, as such approaches can illustrate. Still, it exclusively highlights the inventory that pertains to the problems we find with conventional food chains, highlighting the role of parts only in a limited capacity. While one can use different tools for the same task, say a crescent wrench or a socket wrench, specialized tools provide a way to handle specific tasks more efficiently and or effectively. For dealing with an overabundance of parts, a mereologically inspired theoretical device helps us frame the issue in a way that reveals globalized opacity and its problems, a condition associated with having too many abstract and concrete parts.

During the last chapter, I mentioned that mereology could benefit how we deal with such concerns. This manner of focusing on part-to-part and part-to-whole relations is a quality that is inherent to the mereological enterprise. It is one that "complexity thinking" cannot deliver due to lacking mereology's unique orientation. By examining the composition of food chains by employing a mereological approach, we can exhibit that conventional food chains require expansive infrastructures to exist in their current states. In turn, we can connect this view with other disciplines to advance our knowledge of food chains.[2]

Consider, for instance, that "food narratives" exist to show the elaborate path for foods (and food-like products) to research consumers. In her intense study of food chain routes in *Food routes: Growing bananas in Iceland and other tales from the logistics of eating,* Robyn Metcalfe (2019) explores the numerous parts and their geographical confluence that delivers foods to us. Using a piece of New York pizza as an example, she exhibits the highly involved process that brings all of the ingredients together that are required to produce a single slice (Metcalfe 2019).

While this dish is only one item, Metcalfe's study reveals that today's foods result from a highly sophisticated distribution chain. It demands intense study to understand (which underscores the importance of her book and other works in food studies). Along with her examination, Metcalfe (ibid.) acknowledges that ethical issues emerge that require a separate area of investigation. Yet, with so many parts from across the globe, we cannot see exactly where the problems arise. Due to these conditions, globalized opacity results, making it challenging to see the connection between foods and harm. To gain a better perspective on how this situation manifests, in the following section, I examine how alternative food chains inherently reduce the number of parts, which food activists argue is a necessary condition for alleviating food-related harms and wrongs. Still, it is worth stating upfront that distance is not an inherently bad element in a food chain. On the contrary, distance in some cases could not matter at all, or it could help produce better outcomes. The point that needs to remain at the center of our thinking here is that distance can lean either for or against ethical outcomes when associated with other conditions.

Many food scholars and activists have focused on reducing the distance that foods must travel to reach consumers. Others have given much consideration to different models that seek to improve foods through design. One issue is that, in the scale of food production and distribution, these efforts remain somewhat embryonic, with positive and neglected dimensions. On the one hand, there is little knowledge of how these approaches will fare when tested against fluctuating social, political, or environmental conditions. Having a significant degree of unreliability when stacked against real-world situations casts doubts on their resiliency, which is a hallmark of conventional food chains.

On the other hand, being in a developing state also means that such conditions are not fixed, meaning that we can study and improve them. Such efforts could lead to better-quality food chains that are not riddled with the troubling outcomes examined in Chapter 2. In turn, the section below brings this aspect into view, charting how such measures have advanced and also remain in a state ready to deliver better realities for food chains.

Alternative Food Chains: Towards Reducing Globalized Opacity

During the last few decades, alternatives to conventional food chains have emerged and progressed, viewed as a means to transform food production, distribution, and consumption. Philosophers and scholars have examined several competing and complementary models, making strong cases for how such chains can mitigate the harms mentioned above (Werkheiser and Noll 2014; Epting 2016). Although advocating for particular approaches is beyond this project's scope, the shared grounds often include smaller-scale operations that eliminate several of the actors that exist between the food producer and the food consumer.[3] While it would be overly ambitious to argue that these alternative chains could replace conventional models entirely in the immediate future, supporting such measures counts as steps in that direction to deal with the problem of globalized opacity.[4]

In a more realistic sense, utilizing alternative chains exhibits how possibilities exist to eliminate some of the global parts of conventional food chains, along with the extreme distances that help yield globalized opacity.[5] Although each replacement part will require advanced study and remain site-specific, examples include food outlets such as farmers' markets (Vignali et al. 2006). They also include community sustainable agriculture initiatives (Vasquez et al. 2017). Community and shared gardens also qualify as smaller cases (Barron 2017). In addition to these measures, researchers show how aquaponic operations can ease the demands of marine ecosystems (Goddek et al. 2019). Culinarily challenged individuals can visit farm-to-table restaurants. Although none of these approaches can eliminate dependence on conventional food chains, if consumers were to access them in concert, they would be able to learn how their food chains fit together, one part at a time.

By bringing the parts of a food chain closer together, we can eliminate components such as storage, refrigeration, and long-distance transport inherent to conventional food chains. One could argue that alternative food chains reduce the need for several such parts, which are also the components that consumers lack knowledge of their existence and how they fit into food chains. The point here is that while consumers are probably aware that foods require transportation, they could be unaware of the expansive network of transport needed to move food vast distances, which is increasingly becoming more global (Ahumada and Villalobos 2009).

This condition suggests that replacing them with local actors such as community farmers means that the parts are knowable, a condition that increases food transparency. For instance, Harvey Blatt (2008, vii) argues:

> Most urban shoppers know that food is produced on farms. But most of them do not know what farms, or what kind of farms, or where the farms are, what knowledge or skills are needed in farming, or how farming today bears little resemblance to farming as practiced a hundred years ago.

Through subtracting unknowable and inaccessible parts while adding parts wherein people can learn about the intimate details of food production and distribution, concerns about globalized opacity diminish. Bearing this point in mind, one can argue that improved transparency could yield more knowledge about the immoral practices that turn people against conventional food chains.

Aside from this issue, one could argue that when consumers substitute distant parts with local parts, aiming to develop an alternative chain, they are merely replacing parts that will not significantly impact conventional food chains. This notion implies that such changes are only "cosmetic" and that meaningful improvements will require steps like working *with* the chain while advocating for change through policy initiatives. Over time, such measures will improve consumer knowledge of food chains, inherently reducing the problem of globalized opacity.

This challenge is formidable. The problem is that it does not consider that replacing parts, even by using a piece-meal approach over an extended duration, could have a cumulating impact on conventional foods chains. This aspect suggests that changing a food chain is possible, even though undertaking such a task could take several decades, and there are elements of power and control, as mentioned earlier.

Although the approaches above merely sketch a possible avenue for reducing globalized opacity, they exhibit that eliminating parts or replacing opaque parts—with ones that we could know—could make food chains align with calls for food justice under the appropriate conditions. However, considering that alternative food chains remain embryonic, further research is required. Such efforts not only include the disciplines that could facilitate such realities as engineering, design, architecture, policy, philosophical undertakings, and "applied mereology" in particular also deserve additional study to determine the appropriate pathways forward for such chains. In the section below, I examine some of the immediate areas that would benefit from further research.

Areas for Future Research

While the exposition above reveals the problematic nature resulting from the globalized opacity of conventional food chains, the following steps coming from applied mereology should be to narrow the focus. This idea includes paying attention to specific part-to-part relations, especially in interactions that affect entire food chains. Developing particular research strands is one direction that such efforts could follow. For instance, particular relationships could benefit from further investigation, focusing on significant areas such as production, distribution, and consumption.

On the one hand, issues that pertain to how parts interact with other parts in these three areas deserve attention. Regarding production, we could examine how agricultural technologies, as parts within the sphere of food production, work with other parts that could lead to outcomes that do not raise concerns, such as those described in Chapter 2. For example, studying how energy usage plays a role in agriculture could lead to improvements that result in less anthropogenic environmental degradation, which is an issue with a history of causing harm.

On the other hand, there is a need to investigate how parts within production impact distribution could benefit from additional exploration. The problem of "food miles" is an exemplar. Proponents of local food frequently champion the consumption and production of food that shares proximity, holding that long-range distribution is the primary area of concern (McWilliams 2009). However, researchers argue that growing food in certain regions requires more energy, suggesting that preferences for local food would exacerbate climate change (ibid.). Yet, by analyzing these macro-parts (production) to macro-parts (distribution) to macro-part (consumption) relationships, researchers can identify how they relate to whole food chains. In turn, they can determine which relationships will yield more ethical outputs.

One significant matter here that will require multidisciplinary study is how emerging technologies such as artificial intelligence (AI) could benefit traditional and hydroponic farming. Drones, urban robots, and automated electric vehicles could enhance food delivery. Considering that select cities have some of these technologies in their early stages of development, holding that they could become ubiquitous is not a far-fetched view suited for science fiction. While several philosophers such as Patrick Lin (2012) and

Michael Nagenborg (2020) have given significant attention to how these kinds of technologies will impact human populations, the future will require that these devices become mainstays in our fields and streets, having specialists onboard to guide their ethical implementation and upkeep should accompany these efforts to deliver the world that we want.

This point suggests that considering the practical nature of this work, philosophers will have to engage in interdisciplinary research or work with researchers who are external to the discipline. For instance, philosophers could join researchers from outside their training to assess ethically how food chains fit together ontologically. In this regard, we see the benefits of how philosophers could employ their skills to provide new avenues for exploring the realities behind foods, which is a branch of inquiry that remains neglected.

Despite the practical importance of such endeavors that can help deliver the initial relief that such pursuits can provide, philosophers must also deal with the theoretical aspects that will inform them. Such efforts are required for the long-term restructuring of the mindset vital for fundamentally changing food chains. This point entails examining the base-level thinking about the technologies employed in agribusiness. When this thinking is well-ordered, we can frame it as a kind of technology that we invent to help us complete a task. The task, in this case, is designing and reconfiguring the tools of agribusiness to reduce or eliminate the variety of harms examined in Chapter 2. The following chapter does just that. It starts at the foundations of our thinking regarding international food trade, moving toward advanced classifications of agribusiness technologies that focus on parts and part-to-part relationships as they affect the totality of a food chain. Having fleshed out this view, we will be well-positioned to make ethical evaluations to facilitate desired food chains that yield desired outcomes.

Conclusion

This chapter shows how globalized opacity makes it difficult to comprehend conventional food chains' full scope and impacts, considering that people cannot readily see the sheer number of involved parts. Due to this situation, people cannot understand how food chains play a role in several issues examined in Chapter 2. Although this situation is problematic, by using an applied mereological approach, we can analyze conventional food chains to identify ways to reduce the number of parts that obscure our view of their composition. In turn, researchers searching for ways to eliminate or minimize globalized opacity can develop alternative parts that provide "food transparency" or "globalized transparency."

Although this approach helps us see some of the problematic elements associated with how food chains fit together, numerous research areas on the production, distribution, and consumption of food will require attention. Bearing in mind that the nature of this work is inherently interdisciplinary, philosophers should collaborate with other scholars and activists, revealing how the tools of the philosophical enterprise can provide insights into these affairs. Although such concerted efforts show that issues of food chains demand practical solutions, they also require a palate for the love of wisdom.

Notes

1 To avoid confusion, it is worth mentioning that in the use of the term "part" in this sentence, I am not making a mereological claim per se, but I am dealing with the term in a more ordinary sense of it.

2 This notion gestures towards Carlo Cellucci's notion of a "heuristic view," which seeks to establish criteria for what counts as a fruitful enterprise in philosophy. Namely, to fit such a description, philosophy must be continuous with the sciences, making use of the results that come from scientific discovery, and striving to obtain a perspective that is global. For more information, see Cellucci (2014).

3 This point does not suggest that removing "food miles" is a solution. It only indicates that reducing the number of actors within a food chain could reduce globalized opacity. These issues are connected, but they are not mutually exclusive. Addressing such affairs is far beyond the scope of this book. For more information, see McWilliams (2009).

4 While the primary concern here rests with the issue of globalized opacity, this situation raises additional concerns about the identity and morality of a food chain. That is, would it have a new whole food chain or a food chain with new parts? While either answer requires significant dedication to provide a robust answer, either response does not do away with globalized opacity, meaning that this topic must be dealt with at another time. Secondly, one could argue that eliminating certain actors in the world of global food production might not be fair, considering that doing so could discount the importance of food exportation for developing nations. For more information on this topic, see Navin (2014).

5 This point does dismiss the possibility that local actors could engage in harmful practices, but such affairs could be addressed at the local level when applicable.

References

Ahumada, O., & Villalobos, J. (2009). Application of planning models in the agri-food supply chain: A review. *European Journal of Operational Research*, 196(1), 1–20.

Barron, J. (2017). Community gardening: Cultivating subjectivities, space, and justice. *Local Environment*, 22(9), 1142–1158.

Blatt, H. (2008). *America's food: What you don't know about what you eat*. Cambridge, MA: MIT Press.

Cellucci, C. (2014). Rethinking philosophy, *Philosophia*, 43, 271–288.

Epting, S. (2016). Participatory budgeting and vertical agriculture: A thought experiment in food system reform. *Journal of Agricultural and Environmental Ethics*, 29 (5), 737–748.

Gamboa, G., Kovacic, Z., Di Masso, M., Mingorría, S., Gomiero, T., Rivera-Ferré, M., & Giampietro, M. (2016). The complexity of food chains: Defining relevant attributes and indicators for the evaluation of food supply chains in Spain. *Sustainability*, 8(6), 1–23.

Goddek, S., Joyce, A., Kotzen, B., & Dos-Santos, M. (2019). Aquaponics and global food challenges. In *Aquaponics Food Production Chains* (pp. 3–17). Cham: Springer.

Lin, P. (2012). Introduction to robot ethics. In P. Lin, K. Abney, & G.A. Bekey (Eds.), *Robot ethics: The ethical and social implications of robotics*, Cambridge, MA: MIT Press.

McWilliams, J. (2009). *Just food: Where locavores get it wrong and how we can truly eat responsibly*. New York: Little, Brown and Company.

Metcalfe, R. (2019). *Food routes: Growing bananas in Iceland and other tales from the logistics of eating*. Cambridge, MA: MIT Press.

Nagenborg, M. (2020). Urban robotics and responsible urban innovation. *Ethics and Information Technology*, 22(4), 345–355.

Navin, M. (2014). Local food and international ethics. *Journal of Agricultural and Environmental Ethics*, 27, 349–368.

Vasquez, A., Sherwood, N.E., Larson, N., & Story, M. (2017). Community-supported agriculture as a dietary and health improvement strategy: A narrative review. *Journal of the Academy of Nutrition and Dietetics*, 117(1), 83–94.

Vignali, C., Guthrie, J., Guthrie, A., Lawson, R., & Cameron, A. (2006). Farmers' markets: The small business counter-revolution in food production and retailing. *British Food Journal* 108(7), 560–573.

Werkheiser, I., & Noll, S. (2014). From food justice to a tool of the status quo: Three sub- movements within local food. *Journal of Agricultural and Environmental Ethics*, 27(2), 201–210.

5 Agribusiness Technology

Introduction

Agribusiness technologies were designed and built following principles and complex systems of thought, from mathematics to programming languages and beyond. Yet, the thought systems that guided them were also technologies. Taking this point seriously means that if we can criticize the outcomes that come from the physical parts of food chains, we can do the same with the non-material parts. Fleshing out these notions is this chapter's purpose. It begins by examining Martin Heidegger's conceptions of revealing and enframing technology, showing that the outcomes associated with today's agricultural technologies exceed his descriptions (Epting 2021).[1]

Additionally, the technologies required to save us from our demise demand agricultural technologies that differ from the categories associated with harms discussed in Chapter 2. While examining these points inherently benefits how we understand technology's progression via distinct categories, the primary motivations extend beyond that enterprise. The first goal is to use such accounts to identify the significance behind different agricultural technologies. The second is to provide a platform for making ethical assessments that arise from how these technologies are socio-politically and ecologically situated. After establishing this position, the next chapter moves in that direction.

A Taxonomy of Agribusiness Technologies

Heidegger grounds his thinking about kinds of technology in ancient philosophy, appealing to root-word origins and how they inform his thinking. This move helps him in two ways. First, it exhibits what is happening "behind the scenes" when employing technology to complete a task. Second, as people who can differentiate categories of technologies, it shows why it holds significance. For instance, Heidegger argues (1977/1993, 318):

> Technology is a way of revealing. If we give heed to this, then another whole realm for the essence of technology will open itself up to us. It is

DOI: 10.4324/9781003255505-5

> the realm of revealing, i.e., of truth. This prospect strikes us as strange. Indeed, it should do so, as persistently as possible and with so much urgency that we will finally take seriously the simple question of what the name "technology" means. The word stems from the Greek. *Technikon* means that which belongs to *technē*. We must observe two things with respect to the meaning of this word. One is that *technē* is the name not only for the activities and skills of the craftsman but also for the arts of the mind and the fine arts. *Technē* belongs to bringing-forth, to *poiēsis*; it is something poetic.

In the passage above, we see what could qualify as the limits of how we conceive of a technology. For example, by holding that technē extends to the "arts of the mind," we could interpret that the creative force behind thinking can also lead to applicable insights. We need to make a useful distinction here, holding that well-ordered reason moves beyond art to *lean* in the direction of science or mathematics to the extent that "art" is not a fitting term alone. From that point, enhancing this "art" through providing more structure, such as well-ordered thinking, there is enough difference to hold that we have moved from art to technology—even though the passage above makes room for this actuality itself. Still, art can be highly structured, but we are employing philosophy here to secure a sense of accuracy about the pattern of thought pertaining to agricultural technology.

However, the goal is not merely to use well-ordered thinking for the sake of accuracy. Finding a middle ground that is a well-ordered way of thinking inspired by the mind's creative abilities moves in the right direction. The primary goal (and motivation) is not to discover objective knowledge. Instead, it is to develop or discover a way of thinking that can help us make sense of and deal with the world we inhabit. Again, this notion qualifies as a kind of art, but it favors sciences or mathematics' accuracy. Put coarsely: you could refer to it as philosophy or the pursuit of wisdom. Yet, this conversation exceeds the present undertaking. Still, the takeaway here is that technology begins with advanced, well-ordered thinking. It extends from the mind to the physical devices that help us complete agribusiness. It is between art and science, which fits with philosophy (especially when dealing with the "ordered creativity" required to approach agribusiness technology).

Recalling the two points preceding his passage, we see that there is always an insightful narrative when we use a technology. If we unpack it, we will know that it provides insights into the surrounding elements involved in completing the task. For us, this story could include the mind's capabilities to plan extensively with purpose, using a greenhouse to grow lettuce predictively on a schedule or harness the wind's power. Second, we can examine what such actions mean for us. For instance, creating a growing methodological scheme helps show our mind's ability, that we know how to invent and use tools, and, as we will see in the following chapter, that we can assess ourselves by the technologies that we employ and what they do for us and others. This point means that we can work towards gaining a panoramic view of humankind's agricultural efforts, from our abilities to plan extensively about trade to how we market food products for consumers. The pattern behind such undertakings is nothing new, but its particulars manifest uniquely to contemporary humans. For instance, Heidegger (1977/ 1993, 318–319) exhibits how the pattern of this thought runs through Western philosophy's history, highlighting its significance:

> The other thing that we should observe with regard to *technē* is even more important. From earliest times until Plato the word *technē* is linked with the word *epistēmē*. Both words are terms for knowing in the widest sense. They mean to be entirely at home in something, to understand and be expert in it. Such knowing provides an opening up. As an opening up it is a revealing. Aristotle, in a discussion of special importance (Nicomacheun Ethics, Bk. VI, chaps. 3 and 4), distinguishes between *epistēmē*. and *technē* and indeed with respect to what and how they reveal. *Technē* is a mode of *alētheuein*. It reveals whatever does not bring itself forth and does not yet lie here before us, whatever can look and turn out now one way and now another… . Technology is a mode of revealing. Technology comes to presence in the realm where revealing and unconcealment take place, where *alētheia*, truth, happens.

This passage's significance rests in the consideration that being an expert in thinking about something, which one could say is a well-ordered ability to think, remains in the area of being an expert in using a technology, a technology-thinking specialist. For present times, the philosopher fits into at least two categories. One, the analytic, deals with arguments with computer-like precision. The other sort deals with assessing dimensions of the world, hoping to gain wisdom about it, allegedly. They move beyond theory to practice, dealing with the socio-political-material world's discrepancies as they arise. Like a specialized technician who can complete a task that involves highly advanced skills to complete a job, thinking in such a manner accomplishes goals that adhere to a similar pattern.

The task above only reveals what needs to be revealed, what is out of view conceptually. This understanding is philosophy as a technology. The analytic philosopher focuses on exposing what is lurking behind arguments, their logical or illogical conditions. The contemporary "new philosophers" do not disregard such measures, but they employ ways of thinking (including analytic means) to arrive at different ends, which is to gain insight into a state of affairs about the existing world.[2] The former accounts for most mainstream philosophy, and the latter is people striving to make sense of *and deal with* humankind's experiences in the present world. For the case at hand, philosophers of food fit the description. Philosophers mentioned in Chapter 1, such as Paul Thompson, Samantha Noll, and Ian Werkheiser, come to mind. They employ philosophy to reveal aspects of contemporary life to expose hidden elements from the immediate view. Unlike an investigative reporter, though, their insights contain specialized knowledge that informs us about the specific dimensions of the world, such as their ontological, political, or moral natures, ones that require more than simply pointing out the obvious.

Heidegger is engaging in that kind of activity. After establishing his grounding of concepts such as revealing and its significance, he then illustrates the need for additional thinking that can grasp the extent the modern world differs from the ancient one. He uses philosophy as a technology to reveal how the thinking behind modern technology is a specific kind that should come with a warning label. In sum: he uses one technology to warn us about another one, "enframing."

Agricultural technologies illustrate this point exceptionally well. Heidegger (1977/1993) uses farming technologies to show how they reflect more expansive thinking practices that pertain to how humankind develops and maintains the ability to hold beings and things in different regards. For example, he examines ancient farming practices such as plowing a tract of land to grow vegetables. This kind of technology "reveals" that our thinking can discover how simple measures can provide food from the elementary workings of oxen, plow, and natural elements such as soil, seeds, and water. In turn, Heidegger calls these kinds of technologies "revealing" ones. They reveal what is always there when it comes to these combinations.

For Heidegger, modern technology markedly differs. For instance, it still "reveals" powers that lay hidden, exposed by ingenuity and elbow grease. However, the manner wherein they surface includes an additional element absent from earlier forms of technology. As Heidegger (1977/1993, 320) puts it:

> What is modern technology? It too is a revealing. Only when we allow our attention to rest on this fundamental characteristic does that which is new in modern technology show itself to us... .[Y]et, the revealing that holds sway throughout modem technology does not unfold into a bringing-forth in the sense of *poiēsis*. The revealing that rules in modern technology is a challenging [*Herausfordern*], which puts to nature the unreasonable demand that it supply energy which can be extracted and stored as such.

When he points out we focus on this essential feature of modern technology, Heidegger uses philosophical examination as a tool to reveal the essence of modern technology. This "unreasonable demand that it supplies energy" is the vital element separating ancient and modern technology. As he clarifies, this dimension holds something, typically from the nonhuman world, in "standing reserve," which is required to maintain the operation (Heidegger 1977/1993). These technologies are called "enframing" ones.

Think about how a dam on a river holds the water in standing reserve to give us energy. As a modern technology, the dam and the river can provide electricity, which becomes revealed as something that was "there." The primary difference between these kinds of technologies is the standing reserve associated with enframing technologies. Modern agricultural productions such as factory farms or monocropping with certain fertilizers illustrate how animals or land can be held in standing reserves, counting as enframing technologies. When considering the treatment of many food laborers, one could argue that they receive the same treatment, from production to distribution, as explored in Chapter 2.

These points require us to face the reality of Heidegger's thinking on enframing. Although there is a significant worry for the nonhuman world, such concerns are secondary. The primary fear is that thinking in this manner will become the only way humans think and relate, meaning that we will look at each other only in the same way, to be held in standing reserve. Despite the reality that the corporate world already embraces the idea of humans as standing reserve with gusto, arguably, there is a powerful argument that humankind has not entirely reached the point of no return. This notion suggests that humanity is not ultimately doomed. Still, much damage has accumulated his Heidegger fleshed out his thoughts on enframing.

While these ideas are rather gloomy, the central notion worth underscoring is not that we can pretend that we can avoid technology, claim that we should not depend on technological fixes, or wish that adjustments to human behavior will save us. Instead, we should strive to form "free relationships" with it. In turn, we can enframe or just reveal if we desire, even though we should remain judicious when it comes to the former. Both enterprises stem from our creative powers. As Heidegger (1977/1993, 340) puts it: "But where the danger is, grows The saving power also…" This notion highlights that our capacities to engage in practices (including thinking) that reduce people to things—or think in ways that portray humans as beings who cannot see the full value of things—can lead us out of harm's way.

One theme in the philosophy-of-technology literature is that modern technologies differ drastically from ancient ones, which emblematizes humankind's means of relating to ourselves and the nonhuman world (Jonas 1984). In turn, to see the "full value of things" means not only to provide some consideration for beings and artifacts external to us, but it also entails that we bring a much broader conception into our thinking. Taking this approach seriously requires looking beyond the immediate measures to provide needed relief.

For instance, in thinking about modern agricultural practices, it would be naïve to believe that simply advocating for increased organic measures would suffice. Even though such actions are supportive, today's challenging conditions associated with climate change force us to deal with complicated matters that extend further than simple mitigatory efforts. This point hints at a significant element missing from Heidegger's conception of enframing. The feature that is absent from his description of enframing, which cannot be denied when bringing climate change into the picture, is that such technologies play roles *in* climate change. Considered as part of the totality of agriculture, the associated technologies are no exception, considering that they account for 25–30 percent of global CO2 emissions (Gladek et al. 2017). Of course, there are also the numerous effects discussed in Chapter 2.

In my last book, *Saving Cities*, I introduced the idea that we now need to go beyond enframing to adequately account for modern technologies' role in climate change, focusing on their impacts as a "wicked problem" (Epting 2021). Developed as a conceptual device for urban planning, "wicked problem" has been beneficial for dealing with numerous kinds of problems. Two philosophers working on food issues, Paul Thompson and Kyle Whyte (2011), painstakingly illustrate that this term is incredibly helpful in helping us wrap our heads around value-riddled agricultural issues, including its significance for climate change. Through tracing the term's use across the academy, they show its wide-scale applicability for situating matters in a way that lets researchers grasp their nature, which provides a clear perspective of such problems. Many literature reviews that exhibit a wicked problem's distinct features make this notion apparent. For example, Peters (2017, 388) summarizes them as follows:

> (1) Wicked problems are difficult to define. There is no definite formulation. (2) Wicked problems have no stopping rule. (3) Solutions to wicked problems are not true or false, but good or bad. (4) There is no immediate or ultimate test for solutions (5) All attempts to solutions have effects that may not be reversible or forgettable. (6) These problems have no clear solution, and perhaps not even a set of possible solutions. (7) Every wicked problem is essentially unique. (8) Every wicked problem may be a symptom of another problem. (9) There are multiple explanations for the wicked problem. (10) The planner (policymaker) has no right to be wrong.

Thinking about agriculture's role in climate change and its alignment with wicked problems described above gives reasons to name them "wicked technologies" as a shorthand. The new conditions linked to their ubiquitous nature underscores the significance of studying the link between agribusiness and climate change's impact on the experienced world. Giving this point adequate attention means addressing it at its foundation. That is to say, recalling a point from above, we are essentially dealing with the thinking behind agricultural technology and agribusiness. Our thinking on such matters, which can be seen as a technology itself, as a well-ordered enterprise, also qualifies as a wicked technology. This condition is because we cannot divorce it from outcomes discussed in Chapter 2, as they align with the characteristics of a wicked problem, despite the positive results listed in Chapter 3. This point does not entail ignoring the issues listed in Chapter 2, but such matters often coextensively align with the description of a wicked problem. Perpetuating the harms of the technological status quo connected to this kind of thinking is the foundation for all guided actions. Failing to improve it seems unlikely to deliver a future that significantly deviates from the outcomes associated with climate change.

We must start with how we conceptualize the role of our thinking for how a technology impacts humankind and the nonhuman world, along with all relevant stakeholders. Engaging in such a practice could save us from an anthropogenic demise. However, it also saves us from solidifying ourselves as beings who only think in a manner that sees the world only as a standing reserve. Bear in mind that forming a "free relationship" with technology does not exclude the possibility of employing enframing technologies, but it means that we have the option to do so when the need arises. For example, in a (real) world where binaries are elusive, embracing a choice that involves a standing reserve, such as harvesting biomass for fuel, becomes the best "worst-case" scenario that we can defend.

While this point is crucial, it only shows the surface of a much more critical issue that requires elucidation. Going beyond this notion will also reveal better technological futures, especially agricultural ones. For instance, by engaging in saving thinking, which is also a technology as a well-ordered reason that can help create a better world that reduces enframing or fosters free relationships with technology, we facilitate the process of seeing additional aspects of what humanity can become. That is, by discovering, supporting, or implementing practices across the entire food chain that work to "save" humankind from an anthropogenic demise and reflecting a better way to interact with the world, we create saving habits that help define ourselves and our roles in it.

Moreover, suppose we are to entertain the notion that much of the corporate world, including agribusiness, has already reduced humanity to nothing more than "parts" of the food supply chain. In that case, we can mitigate the harms embedded in such practices. With this point in mind, we see that we are not merely talking about planning for the future, but we are making efforts to address problems that exist at present, from the farm to the table. The significance of this point rests on the idea that saving technologies should do more than mitigate the harms associated with climate change. They should also engage with the full range of harms and values associated with wicked technologies, which requires examining numerous kinds of impacts on the world.

Considering that we are dealing with food, the argument that we should depend less on technology is extremely weak, especially when facing the reality that we are talking about human populations across the world. The question that needs formulating is not how can we move away from wicked technologies to deliver better food outcomes. Instead, how can we shape questions about agribusiness that can maximize the utility of outcomes that stem from sophisticated operations in ways that respect a wide range of stakeholders?

To investigate arrangements in agribusiness on fundamental levels, as established in the last chapter, sticking with the "parts" concept continues to provide benefits. It allows us to see interactions between technologies as they yield specific results that we can analyze in subsequent chapters. Considering that harms come from arrangements of agricultural parts, lack of parts, and or control over parts, focusing on these aspects reveals troubles on several levels, from local to global and back. It also makes the process of moving food business towards a more ethical status more manageable. In turn, identifying problems using this approach sets the stage for making moral assessments of parts' arrangements within larger food systems in which they are found. This point leads us to the next chapter's focus. It deals with competing interests belonging to the kinds of stakeholders discussed in Chapter 2, which include marginalized groups, exploited workers, harmed consumers, nonhumans, distant generations, and the artificial world in some instances.

Areas of Future Research

Although dealing with arrangements of agricultural parts can benefit the task of identifying harmful configurations, the significant physical challenges rest with engineers, scientists, and logistics professionals. They are the people who can provide relief to many troubling situations, such as those in Chapter 2. One significant benefit of Metcalfe's (2019) *Food routes* is that she details how numerous areas in food logistics are steadily advancing to bolster food chains, from artificial intelligence (AI) to storage, genetic modification, shipping, storage, and beyond. While many of the problematic issues found in food chains might appear to be collateral damage in battles for ground in the global marketplace, digital and physical technologies could help decrease tensions associated with agribusiness. This point suggests that we cannot discount the importance of pursuing new or improved agricultural technologies across food chains.

However, their monumental significance does not dismiss or undermine philosophers' contribution to working on wicked problems and securing the involvement of saving technologies. When it comes to wicked problems in agriculture, Thompson and Whyte (2011, 494–495) make this notion apparent:

> [T]he most relevant contribution that philosophers can make will very likely depend specifically on the wicked problem context… . In analyzing wicked debates over the future of genetically engineer crops, it has been more relevant to discuss how the partisans in the debate over labeling have dismissed consent-oriented claims with utilitarian arguments that beg the relevant ethical questions (Thompson 2002) or how competing analyses of the environmental risk from these crops have made subtly different epistemic assumptions in the construction of their

> risk analysis (Thompson 2003)… .Given any epistemic orientation consistent with problem-responsive inquiry, the wicked problems framing will allow participants from a wide variety of backgrounds, knowledge orientations and capabilities to work collaboratively. In this, the skills that philosophers already have can be emphasized in interactions with others and turn out to be valuable to team-based learning processes and strategies for addressing wicked problems.

The passage above conveys the vital idea that dealing with agribusiness reforms must be an interdisciplinary (or multidisciplinary) affair. While this approach might offend the disciplinary sensitivities of orthodox academics, it aims to meet the problems identified in Chapter 2, where they are conceptually. This point is not meant to attack purists, but it does account for a full view of the necessary positioning to advance reform efforts optimally. The pattern behind this view has been established in many philosophical literatures, suggesting that the future of philosophical support for agricultural reform will continue to evolve. In turn, it will move in the direction that Thompson and Whyte (2011) predict. One benefit unique to philosophy is that the "new" philosophy typically crosses external disciplines and areas of study. The point suggests that one contribution that philosophers can make is navigating between disciplinary outlooks to make connections, paying close attention to congruence or disconnects between theoretical devices.

For example, in the philosophy of food, Samantha Noll's (2015, 2017) work on urban food policy and practices navigates between history, urban planning, and public health, providing significant insights into contemporary affairs in ways that see far beyond the surface of related matters and affairs. Due to her orientation and ability to intersect with these distinct areas, the benefits of understanding the food-city dynamic are enhanced by giving a more precise, more profound understanding of the ontological history of cities' food chains. Giving such studies attention could help us understand the past and the present, but it could also help us avoid repeating the same or similar mistakes. Scholarly engagements such as this one push against philosophy's disciplinary norms, moving it into a new realm that differs from academic orthodoxy.

Consider, for example, Lewis Gordon's (2015) conception of "disciplinary decadence" informs the view that purist philosophers see their positioning as a sacred, timeless form that sits above knowledge. They expect the world to conform to their views instead of adjusting their views to match the world. It is easy to criticize issues because they do not fit within theoretical frameworks that determine fairness, justice, or morality, but they do not start with the situations at hand and deal with conditions found in the world. This point does not suggest that wondering in the armchair is fruitless. Yet, it does show the limits of such exercises. The problem with food-reform efforts is that philosophical insights rely on bodies of knowledge than other disciplines produce. That is, philosophical contributions such as mereology and ethics are required to make sense of agribusiness (and the world). For such cases, employing such thought technologies must be done in concert with other tools of the academic trade. These devices include but are not limited to statistical analysis, data sets, engineering principles and practices, and narratives.

In turn, the future contribution from philosophy must act *with* these bodies of knowledge instead of being above them. A purist could push back against this point, holding that all bodies of knowledge come from epistemological groundings that philosophy provides. This debt is oversold considerably. While this point is not wrong per se, it grossly overestimates its significance. It arrogantly assumes that scholars outside of philosophy are incapable of thinking logically without guidance from theoretical overlords. Such attitudes are what Robert Frodeman and Adam Briggle (2016a, 2016b) have warned philosophers about, that the discipline has produced isolated intellectuals who cannot find purpose in the academy beyond the importance that they assign to themselves.

Moving forward, philosophical insights into agribusiness reforms must join interdisciplinary conversations wherein they have something to offer that complements other scholars' bodies of knowledge. That outcome has been the goal when dealing with the ontological structure of agribusiness, as discussed in the previous chapters. It continues by providing ethical assessments in the next chapter. The point here is not to only argue that the actions addressed in Chapter 2 are wrong. That point is obvious. Instead, it looks at the immoral enmeshment that is those problems, arguing that the manner wherein we approach them is itself a moral affair, a sophisticated incidence of the problem of moral prioritization. The issue here is that when confronting a case involving several stakeholders, the group that receives the first attention is an ethical matter itself.

For instance, Chapter 2 identifies several issues that affect or will impact people, ecosystems, and artifacts (in some cases). While a universal-care approach would be optimal, the world does not often present such situations. In turn, "meeting the world" means acting for one group first. This chapter takes on this challenge, unapologetically holding that suffering people deserve it.

Conclusion

This chapter examined Heidegger's notions of revealing and enframing in the context of agribusiness. After gaining clarity on the benefit of employing these concepts, the attention turned to how they are ill-equipped to categorize food technologies that contribute to wicked problems such as widespread social inequality and global climate change. To account for such conditions, I showed how introducing the category of "wicked technologies," inspired by the thinking behind wicked problems, can extend how we categorize kinds of technologies that now have a wicked problem strongly associated with their existence within more extensive socio-material-environmental arrangements.

Instead of merely aiming to identify additional problems, I then illustrated how moving toward "saving technologies" provides a way out, moving away from wicked and, when fitting, enframing technologies. Attacking this problem at its foundation means examining such affairs at their base, meaning that we must shift or create a well-ordered way of thinking about agribusiness that can yield outcomes that can save us from our peril. While such efforts could help advance conversations about what is required to gear food-reform efforts in a beneficial direction, many additional advancements are needed.

On the one hand, scientists, engineers, and logistics specialists deliver and will continue to provide technological solutions that mitigate select harms, as discussed earlier. On the other hand, such matters will also require the assistance of philosophers. They can navigate interdisciplinary waters, provide ontological assessment, and deal with the inherently ethical nature of agribusiness, which also challenges what it means to do philosophy.

Notes

1 I also examined many of the points in this chapter in relation to urban technologies. See Epting (2021).

2 While I am calling such enterprises "new philosophy," the reality is that such endeavors are anything but new. The history of philosophy is filled with people engaging in this kind of work. However, the adjectives that would work better already have uses that do not exactly correlate to such terms. For example, "applied" philosophy, one could argue, is basically just standard philosophy that slightly differs, and it does not describe the kind of work in question. However, this topic falls beyond this book's scope, meaning that it is better suited for future investigations.

References

Epting, S. (2021). *Saving cities: A taxonomy of urban technologies*. Cham: Springer.

Frodeman, R., & Briggle, A. (2016a). When philosophy lost its way. *The New York Times*, January 11.

Frodeman, R., & Briggle, A. (2016b). *Socrates tenured: The institutions of 21st-century philosophy*. London: Rowman & Littlefield International.

Gladek, E., Fraser, M., Roemers, G., Sabag Munoz, O., Hirsch, P., & Kennedy, E. (2017). The global food chain: an analysis, *Metabolic*. Available online: https://www.metabolic.nl/publications/global-food-chain-an-analysis-pdf/.

Gordon, L. (2015). *Disciplinary decadence: Living thought in trying times*. New York: Routledge.

Heidegger, M. (1977/1993). The question concerning technology. In *Martin Heidegger: Basic Writings*. Edited by D. Krell (pp. 311–341). New York: HarperCollins.

Jonas H. (1984). *The imperative of responsibility. In search of an ethics for the technological age*. Chicago: Chicago University Press.

Metcalfe, R. (2019). *Food routes: Growing bananas in Iceland and other tales from the logistics of eating*. Cambridge, MA: MIT Press.

Noll, S. (2015). History lessons: What urban environmental ethics can learn from nineteenth century cities? *Journal of Agricultural and Environmental Ethics*, 28(1), 143–159.

Noll, S. (2017). Food sovereignty in the city: Challenging historical barriers to food justice. In *Food justice in US and global contexts* (pp. 95–111). Cham: Springer.

Peters, B. (2017) What is so wicked about wicked problems? A conceptual analysis and a research program. *Policy and Society*, 36 (3): 385–396.

Thompson, P.B. (2002). Why food biotechnology needs an opt out. In B. Bailey & M. Lappe (Eds.), *Engineering the farm: Social and ethical aspects of agricultural biotechnology* (pp. 27–44). Washington, DC: Island Press.

Thompson, P.B. (2003). Value judgments and risk comparisons: The case of genetically engineered crops. *Plant Physiology*, 132, 10–16.

Thompson, P.B., & Whyte, K.P. (2012). What happens to environmental philosophy in a wicked world? *Journal of Agricultural and Environmental Ethics*, 25(4), 485–498.

6 Moral Ordering for Agribusiness

Introduction

So far, much of the attention has focused on modern technology's non-normative roles in agribusiness. It starts with how we think about food production and distribution, permeating almost all aspects of those processes. Yet, the focus remains mainly on the food chain's composition, breaking it down to look at how parts fit and work together to yield a mixed bag of good and bad outcomes. As argued, moral progress does not keep pace with technology's advancement (Tremblay 2010). This point signals that we are shifting from concerns about how agribusiness is in the world to how it should be ethically. As explored in Chapter 2, there are numerous moral issues enmeshed within agribusiness. This chapter aims to move in this direction, providing a way to address the many *kinds* of problems in tandem. This point means that we must confront the problem of moral prioritization, as mentioned earlier.

In turn, this chapter begins by examining Hans Jonas' critique of canonical moral theory, attending to the intersection of humankind's permanence, technology, and the nonhuman world, focusing on how agriculture fits into the picture. Having established this view, I turn to a novel and emerging approach to moral extension-ism, moral ordering. This process aims to lessen the issue of moral prioritization in multi-stakeholder engagement. With an understanding of this measure laid out, the chapter ends by highlighting some of the necessary steps that could help execute measures essential to usher in agribusiness reforms.

The Need to Expand Ethical Coverage to Agribusiness

While there are still a low number of philosophers working on food issues compared to mainstream areas, philosophical literature has made significant contributions to providing clear thinking about agriculture and food. In addition to the contemporary philosophers mentioned in the previous chapter, Hans Jonas (1984) has wrestled with the intersections of morality, humankind, nonhumans, technology in general, and, to a lesser extent,

DOI: 10.4324/9781003255505-6

specific agricultural technologies during the 1970s. Writing long before environmental ethics was an established subfield within academic philosophy, his views now seem prescient, especially for modern farming practices (Epting 2010). In particular, he focused on how specific farming applications played a role in threatening our species (Jonas 1984).

During the 1970s and 1980s, Jonas engaged in philosophical pursuits that largely avoided questions about technology's ontological status, opting to examine how it produced outcomes that required advanced moral study instead. However, canonical moral theory lacked the necessary elements to make moral sense of modern technology. Ideas from figures such as Aristotle, Immanuel Kant, John Stuart Mill, and Jeremy Bentham had several limitations in this regard. The primary issue here is not that their approaches were fundamentally flawed. Jonas argues that these theories were made to address immediate anthropocentric matters, and they still can hold their own when dealing with such issues.

However, when going beyond the parameters of those kinds of affairs, they lack the vital theoretical components to get the job done. As technologies of the mind, they were the wrong tools and could not help us make the kind of sense required to remedy humankind's ill conditions. Specifically, they could not account for conditions such as moral consideration for the nonhuman world, namely its intrinsic value. Those philosophers made no room for events that were not immediately known. The matters on their minds were events such as lying, stealing, or murder. The canon did not consider the passing of time's role in humankind's affairs. Although this point might sound trivial, it introduces elements of paramount importance for agribusiness.

For example, the long-term application of some chemical fertilizers and pesticides, which has now had a noticeable impact on the nonhuman world, goes beyond farmlands to include harm to the oceans (De Sylva et al. 2000). The essential point is that established moral theories were not designed to anticipate technology's long-term, accumulating effects on populations and the nonhuman world. In a Jonasian view, modern technology challenged the entire canon of moral thought to the degree that warranted measures beyond merely designing another theory. Instead, Jonas (1984, 11) provides a blanket imperative to govern all technological undertakings: "Act so that the effects of your action are compatible with the permanence of genuine human life." While his criticism is bold and sweeping compared to the traditional objections associated with specific moral theories, much work remains concerning implementing his dictum in particular fields requiring it, such as agribusiness. Recalling a point from the last chapter, much of the fieldwork requires scientists, engineers, and logistics specialists to develop mitigatory efforts. Yet, philosophy must match the pace of their actions.

Addressing this concern at its foundations, we can take this view an additional step, holding those well-constructed moral theories mentioned above are technologies that cannot handle the wicked problems associated with agribusiness. Jonas recognized the need to go beyond those theories' parameters. Since then, numerous philosophers have engaged in moral extensionism, creating approaches to deal with morality and nonhuman species, ecosystems, and technology. In turn, the specific subfields in philosophy have inched their way towards addressing Jonas' concerns. "Moral ordering," a method that can address issues that involve several stakeholder groups, provides an optimal way to deal with moral considerations that speak to Jonas' contentions (Epting 2021). They also offer a way to show that competing interests between stakeholder groups can be complementary.

When seen in a suggestive order, the notion of competition between groups essentially dissolves, providing a clearer view of what it means to exist alongside external kinds. The goal is not to argue that one stakeholder group has a unique status per se as an end in itself. Instead, the motivation is to elucidate that these stakeholder groups require consideration in a certain way and a particular order. As mentioned above, these aspects are suggestive only. This quality accounts for the unique circumstances surrounding issues such as agribusiness. As a technology, moral ordering is adaptive to deal with problems in various contexts (Epting 2021). Even though, as a process, it maintains its basic structure across different situations, its characteristics must change to approach unique scenarios.

Not all stakeholder groups will exist in each situation. The stakeholder groups for a problem in waste-water management might differ significantly from the groups dealing with issues in information-communication technology management. Each kind of problem requires a customized approach to deliver the desired outcomes. In turn, cases in agribusiness need the same requirement, and the sections below provide it.

Towards Moral Ordering in Agribusiness

The points in previous chapters culminate in the following pages, illustrating how we ought to develop a future for agribusiness that differs ethically from the past and present. Bringing this picture into view requires addressing two points in tandem. First, we need to see that the "ought" above is normative. We are not only dealing with a topic that concerns morality, but the manner wherein we deal with the topic is also a moral affair. Put simply: deciding how to approach ethical issues in the food business is a moral issue itself. I have dealt with similar cases elsewhere, referring to the pattern behind such cases as the "problem of moral prioritization" (Epting 2021). Essentially, a meta-level concern lurks behind the decision-making process, asking, "Are we acting morally when making decisions with moral aspects?" The worry here is that ignoring how decisions are made could count as immoral actions, implying a proper way to address such choices.

This issue leads to the second point. That is, the "who" who determines how such matters are dealt with must also hold steady as an element inherent to any such decisions that lead to policy or any similar efforts to reform the moral dimensions of agribusiness. This point must be seen in tandem with the previous one because it concerns how decisions in agribusiness ought to be made. For instance, policymakers or people in positions of power can no longer simply make choices, assuming that the ethical aspects are sufficient. Instead, the people affected by such decisions must have a meaningful way to help determine them. Addressing these points in such a manner favors moral rightness on two fronts.

First, it promotes the idea that morally reforming agribusiness norms and practices will receive the kind of foundational attention that can secure positive changes that become entrenched in the production and distribution of food. The significance of this notion rests on the idea that supply chains must remain constant and predictable, but they must also adapt quickly to changes inherent to such systems. If we recall, some of the harms that field workers endure, which we can call unethical, are routine aspects found in agribusiness. It should not be a stretch to imagine that the same working scenario can deliver better, more ethical outcomes for laborers. Positive change needs to happen, and the pattern behind a supply chain's "behavior" exhibits that disruptions or alterations are typically short-term problems. Adjustments come quickly in many cases (Metcalfe 2019).

Despite a food supply chain's resilient nature, arguments exist against their disruption, and the ugly pattern that replicates them can be found in similar forms. For instance, slavery in the United States was once defended because the economy would suffer if the horrid practice were eliminated (Hampton 2015). *I am not saying that all unethical food production and distribution practices are equal to slavery*—even though some *actually* are cases of slavery and forced labor.[1] Still, many of them are within the scope of ill-treatments that workers endure. Arguments that defend horrid working conditions are not in short supply, and their defenders make them with a straight face. This point aside, the idea here is that people throughout history have defended the oppression of other people when working conditions receive moral evaluations. That trend will continue, unfortunately. What has not been a trend is ensuring that exploited individuals have a feasible way to eliminate harmful conditions associated with their jobs.

The reality that food laborers have endured suffering gives them an unparalleled buy-in to have reform efforts geared in their direction, including meaningfully engaging in such actions. This notion might sound radical, but in the face of suffering and exploitation, getting at the root of the problem that makes ill-treatment possible requires mitigating existing harm, working towards amends for previous injustices, and developing a way forward that is inherently resistant to future abuses. Bearing these points in mind, the next step is to re-examine the range of stakeholders, beginning with those who endure significant harm due to the buy-in that such conditions create. These people are the first group of stakeholders that require a change in food chains that should ease their burdens and benefit their lives. However, each group deserves attention and action.

This process, called "moral ordering," provides a suggestive manner of dealing with such sophisticated affairs inherent to the global food industry. The section below continues fleshing out this order. It illustrates how it is optimal under most circumstances to view complex issues involving several groups of stakeholders. While they seem to be competing for their respective interests, putting them at odds with each other, moral ordering shows how this belief is shortsighted. After unpacking this view, looking at how a food supply chain's parts fit and work together provides a panoramic picture showing a "moral order of operations" required to improve agribusiness ethically.

Moral Ordering in Agribusiness

As shown above, the harms that many food workers have endured provide strong reasons to have their interests firmly in view when thinking about reforming agribusiness. Although the degree of injury varies from one case to the next, the notion that binds them together is that such ill-treatment brings up the idea that amends are required, such as restorative efforts. One lesson from restorative justice is that all parties should work towards making things square between all parties. This point means that they need to examine previous events and create anticipatory measures to ensure that abuses and harms do not happen again.

In turn, seriously considering this notion entails accepting the proclamation of groups such as La Via Campesina (2007) puts forth in the *Declaration of Nyéléni*, which means that policymakers ought to engage with such groups to determine a path forward. Recalling from Chapter 2, the items called for in the *Declaration* are expansive. They touch on aspects that seem only remotely connected to agribusiness. Researchers have pointed out that asking for such a wide array of changes could backfire, making it extra challenging to secure change (Flora 2011; Bernstein 2014; Whyte 2016). While this point is sound, it does not do away with the notion that food impacts those areas. This notion suggests that changes in agribusiness extend beyond immediate issues associated with food, but those affairs have ramifications that people disconnected from food production and distribution cannot see. Due to this condition, people outside agribusiness cannot understand the linkage. Still, this situation does not mean that such issues are not concerns for workers, bosses, owners, or consumers.

However, by bringing those concerns into view firsthand, outsiders could gain the needed perspective to see such connections. Still, seeing the associations is not an essential requirement per se. That view supports a power imbalance that feeds and justifies the business's domination over workers, which, one could argue, perpetuates harm. Unfortunately, exploring this avenue fully involves taking up debates that fall far outside of agribusiness that examine the roots of labor, capital, and trade. The study at hand *depends* on business and the many steps and exchanges that facilitate it, meaning that we are locked into specific kinds of arrangements.

While this situation indicates that we must think within predetermined trade parameters, those boundaries remain fluid and customizable. This idea suggests that almost any substructure of policies, practices, or methodologies remain compatible within the larger agribusiness structure. In turn, such possibilities open up opportunities to level the "ideological playing field" so that a school of thought external to globalized food trade can exist within the more extensive operation of exchanging food for currency. This reality cannot be wished away. Although these conditions create somewhat fixed conditions for agribusiness, they do not take away from the ability to deal with morally rearranging the thinking that governs it. Instead, they *bolster* the ability to establish an ethics of agribusiness that is meaningful, inclusive, adaptive, and expansive. It must include several stakeholder groups that operate within the bounds of agribusiness.

For instance, by calling into question the plasticity of morality's preconditions that come from employing moral ordering in agribusiness and its *suggestive* nature, we open up the possibility to determine the critical elements that guide the outcomes that are objectionable or that we can embrace as ethically sound. Moral ordering, then, serves as the mechanism that makes agribusiness morally palatable. In turn, the endeavor is to get it right for the groups of stakeholders, providing them with the degree of attention that a good moral order can deliver.

The beginning of the section above shows how stakeholders who labor in agribusiness have substantial stakes that demand initial action due to the conditions that current practices have created and perpetuated. Along with these people, consumers are also harmed. Often, they are trapped within the confines of socio-economic, material surroundings. For instance, Julie Guthman (2011) in *Weighing in: Obesity, food justice, and the limits of capitalism* illustrates how health concerns such as obesity intersect with built environments. When it comes to practical and affordable access to healthy foods, numerous groups remain at a disadvantage. They are unable to secure food that can lead to good health.

Terms such as "food deserts" and "food mirages" serve as conceptual devices that give us a way to think about how arrangements of food-system parts help create harmful scenarios from which people cannot escape. Socio-economically disadvantaged groups and others must contend with agribusiness conditions in concert with external systems such as urban planning. The outcomes require identifying, scrutinizing, and then changing agribusiness so that these people have inherently better options that take the same effort as selecting foods that harm. One could object to this point, holding that their agency is disrespected because they do not choose what their options are going to be instead of having better choices. However, this point assumes that people want to exercise agency in such a manner.

The reality is that many foods appear to be healthy based on heuristics, such as food pyramids or food plates. Still, one could argue that deciphering nutritional information requires that people have a complex understanding of such issues. Unless they take it upon themselves to develop the interest to pursue this education, they will not have such knowledge. This point is significant, especially considering that primary schools in numerous locations affecting innumerable people do not provide such lessons alongside staple subjects such as math and science. Although the educational aspects are external to agribusiness, this condition does not dismiss the fact that numerous people's food options are predetermined by multiple parts of agribusiness working to create the conditions wherein consumers are predisposed to select unhealthy foods.

Look at the expansive array of professionals who play roles in creating such outcomes: taste engineers, food psychologists, and marketing executives are dedicated to getting customers to purchase their products. This reality shows that much effort goes into the harmful outcomes for the public. In turn, holding that the typical food consumer can begin to exercise their agency places unrealistic demands on the ordinary person simply dealing with their daily grinds of life. Replacing parts of the food system that currently harm them without bringing agency into question makes this notion evident. Consider, for instance, fats in foods. One could make a case that processed food production could use healthy fats in select products, but it is cheaper to use unhealthy ones. Many foods, such as salads that appear healthy, are the opposite (i.e., fast-food). Removing these harmful parts and replacing them with more nutritious versions would not impact people's agency. It remains the same.

Still, the point worth keeping in view is that people have been harmed on both far ends of food chains. These people require the primary consideration in the grand scheme of agribusiness reform. Again, having been harmed is the condition that gives them buy-in for consideration followed by action. Although this reason holds significance, other stakeholders such as nonhumans also have been "harmed"—at least from an anthropocentric perspective, which is the only fully epistemically ground-able perspective. This latter point entails that, despite advances in understanding how some animals' minds work, we will never know precisely what it means for an animal to have complex interests that can exist external to our anthropocentric interpretation. It should not take a sophisticated argument to show that factory farming is incredibly problematic, as many philosophers argue, to say the least (e.g., Kemmerer 2015).[2] While claiming that we can know what is in the interest of one or a few species is troubling, holding that we can know what is in an ecosystem's interest is far more complicated, considering that ecosystems can only be understood from an anthropocentric perspective (Hargrove 1992). Still, these notions do not justify the complete domination of the nonhuman world for agricultural purposes without holding any reservations for its own sake.

For example, we can be considerate of the nonhuman world's intrinsic value. We assign that value to it when we agree that that world has any such value outside of what it can do for us, its instrumental values.[3] Despite having these values, on the surface, it would appear troubling to say that we must put the nonhuman world's "interests" ahead of humankind's interests—unless there is a compelling reason. Such a move could be equivalent to saying that green lives matter more than human lives, which creates additional complications. For instance, Ramachandra Guha (1989) and Charles Mills (2001) have shown (in a broad sense) that holding such positions creates problematic situations that bring race and geopolitics into view (Epting 2017, 2021). Still, there could be cases wherein we must place nature's "interest" ahead of our own, but it would actually be in our long-term interest to do so.

Considering that—ultimately—all such interests are inherently anthropocentric, placing care for the nonhuman world beneath our direct interest is simply a way to prioritize harms that challenge humankind. For these reasons, the nonhuman world still deserves a place in the moral order, but it must remain beneath our primary place unless exceptional circumstances call for such actions. Although these notions support placing nonhumans beneath humans, the fact that we are talking about agribusiness for the sake of humans inherently positions people before plants and nonhuman animals.

Even though all such actions remain only epistemically grounded on anthropocentric terms does not mean that any such action toward the nonhuman world is morally permissible (e.g., factory farming). Such actions would be saying that "is" equals "ought," which is categorically wrong, a case of the naturalistic fallacy. Instead, recalling from Chapter 3, we require saving technologies to replace wicked and enframing ones to mitigate existing harms and prevent additional ones in the future. This point does not mean that we cannot hold the nonhuman world typically in standing reserve for farming and ranching, but it suggests that we should employ such measures with caution.

A point that must remain in view is that in forming a free relationship with agricultural technology, employing a means of production that involves standing reserves is permissible. However, due to the historical misuse of these practices, determining to engage in such a practice should come with intense scrutiny. In some instances, this kind of technology would support practices that food scholars rally against, such as various forms of farming and biotechnology (Shiva 1993). This notion puts the arrangement of parts back in the spotlight, drawing attention to the technologies that are not inherently good or bad. Instead, parts surrounded by other parts can yield outcomes that fail to pass moral muster. This idea must remain at the center of our thinking about the future of agribusiness. Arguably, failing to have it as a mainstay in food production, distribution, and policy has played a significant role in the horrible outcomes discussed in Chapter 2.

What is more, one could hold that keeping the interconnected nature of human and nonhuman elements will be a continuous challenge. Making room for complicated considerations such as sustainability, which includes future generations, burdens our morally balancing efforts to improve agribusiness.

However, we cannot dismiss the notion that we inherit from Jonas. Technology forces us to consider people who will (hopefully) exist as a moral concern for us in the distant future. Present-day people have to live with agricultural decisions made decades ago, long before we were aware of the enduring effects of farming practices such as the widespread application of pesticides.

As Jonas (1984) puts it in a roundabout way, if we consider that we want there to be people existing in the future who should lead genuine human lives, we owe it to ourselves to secure the means for these conditions. This view puts future generations beneath consideration for the nonhuman world because maintaining that world is required for people's future existence. In turn, we see how future generations need to have this place in the moral order by default, suggesting that agribusiness reforms should keep an appropriate amount of consideration for future people in view. Sophisticated terms such as "sustainable agriculture" have these embedded notions in their definitions.

Lastly, there are many historical tensions between agribusiness and land use globally, internationally, and locally. Recent efforts to reintroduce agriculture into urban environments bring ethical considerations into situations that are already ripe with several other land-use issues. Consider urban farming operations. Many cities are dealing with a lack of affordable housing. Taking up valuable real estate to produce organic lettuce for wealthy urbanites is sure to cause trouble. This point is not meant to argue that urban agriculture should be excluded entirely. Still, it does suggest that such enterprises could require advanced study and debate to identify and deal with the moral parameters of those scenarios.

Along with this issue, concerns such as historic buildings, neighborhoods, and landmark structures (e.g., parks) also must compete for space. In turn, making a case for growing food or raising animals in metropolitan environments could be a hard sell. This last term, "hard sell," means that people require being on board with such endeavors. The point is not to convince them in such cases, but the idea is that they will have to approve of those measures when stacked against other needs. Such a notion is not one that we can cover in passing. The section below examines some considerations that should guide how the implantation of reform efforts in agribusiness intersect with the stakeholder groups above to give it the attention it deserves.

The Scope, Limits, and Promise of Stakeholder Influence in Agribusiness

The points above, coupled with the previous chapters' insights, show the complex moral dimensions enmeshed in the global food marketplace. This broad scope provides innumerable opportunities for acting wrongly and rightly but acting without regard for either could yield higher profits, which speaks to the animalistic nature of business conceived in an extremely narrow sense. While customers concerned with bad actors could speak out against them by not purchasing their products, researchers have made strong cases showing that such measures are short-lived (Lentini 2022). As mentioned in Chapter 3, globalized opacity renders this option useless unless people have the time and desire to research each item relentlessly.

One could posit that businesses should be interested in letting consumers play a role in determining the production and distribution of foods, but this notion makes little sense when considering the competitive nature of business and people's busy lives. Still, agribusinesses that embrace the ethical dimensions outlined above will attract customers who are increasingly showing that they want to purchase goods that have better footprints. Despite the rising popularity of foods that trade on trends such as local, organic, and non-GMO, the reality is that an overwhelming number of people might care less about those features, access them, or cannot afford those products. This point suggests that relying on the market alone to reform agribusiness is highly unlikely. Such an outcome is not guaranteed to account for a significant share of consumed foods in most places. For people who want to live ethically regarding their dietary intake in ways that intersect with the stakeholder groups discussed, employing other measures will have to come into view.

The bad news is that the necessary inroads involve politics, which are highly suspectable to corruption. Yet, the good news is that avenues exist that could facilitate those possibilities, and there are modest cases that provide inspiration and hope for agribusiness reforms. With increasing support and advanced study, bolstering the conditions that make them feasible options could improve the moral dimensions of agribusiness in ways that positively affect the stakeholder groups examined above. The next chapter does just that, showing how several stakeholders can influence agribusiness effectively. It looks at two kinds of cases, one being a "thought experiment" geared to help create saving technologies that can strengthen the ethical integrity of food chains.

Conclusion

After establishing a means to categorize arrangements of parts of food chains in the previous chapters, this one advanced the discussion by providing a way to prioritize the agribusiness issues that require reform morally. This process, a technology exclusively designed for multi-stakeholder engagement, is called moral ordering. It begins with groups that have already been harmed, including but not limited to workers in fields and factories, small-scale farmers, and often consumers. Having already received harm, these unfortunate realities hold the conditions that give them the necessary buy-in to have their interests prioritized ahead of other stakeholders.

Once considerations are in place for such groups, actions should turn towards delivering moral outcomes for the public. This positioning shows that people matter more than nonhumans, but this point does not entail agribusiness can hold nonhumans exclusively in standing reserve. The worry here is that doing so would exacerbate our abilities to deal with wicked problems proactively. In turn, nonhumans deserve moral consideration, which puts them in this position. However, there is no reason why we cannot respect them for their intrinsic value. Still, placing them above humans implies that nonhuman lives matter more, which opens up significant criticisms, especially when dealing with marginalized or historically harmed populations. This point does not exclusively entail that nonhumans can receive special consideration, only that such instances require advanced study to justify their place in the moral order.

Following these positionings, future humans should receive consideration, which is an ethical and practical place. While putting them in this place is a righteous act because we would not want to give these people consideration before those struggling. They have a kind of interest that challenges us to give them attention. Yet, the nonhuman world must survive to have a place to exist in, assuming that they are born. One could also say this group's position in the moral order is mainly determined by default. Lastly, the artifacts such as neighborhoods and buildings should also receive ethical consideration for their intrinsic and instrumental values. This issue could arise in places where cities' needs challenge nearby farming operations.

Although moral ordering appears absolute, it is only suggestive. This point means that if a lower-place stakeholder group has a pressing issue that should weigh heavily, it can move up in the moral order. However, due to corruption and deception, such cases need assessment on a case-by-case basis. Finally, it remains imperative that stakeholder groups can meaningfully co-determine the policies and practices that will shape agribusiness. The vital reform efforts will require significant attention to improve the ethical outcomes that they generate, meaning that saving arrangements of parts will need to be included in food chains or replace bad parts. In turn, the next chapter moves in that direction, examining two different technologies that can help secure such changes.

Notes

1 To learn more about modern slavery and how you can make a difference to stop it, see http://slaveryfootprint.org.
2 As mentioned in Chapter 2, factory farming is a hotly debated topic that also intersects with animal rights. In turn, examining this issue should rely on philosophers and scholars working in that area specifically, which far exceeds the space provided here, considering that it encompasses an entire research strand that has sub-strands.
3 For more information on this position, see Hargrove (1992).

References

Bernstein, H. (2014). Food sovereignty via the 'peasant way': A sceptical view. *Journal of Peasant Studies*, 41, 1031–1063.

De Sylva, D.P., Richards, W.J., Capo, T.R., & Serafy, J.E. (2000). Potential effects of human activities on billfishes (Istiophoridae and Xiphiidae) in the western Atlantic Ocean. *Bulletin of marine science*, 66 (1), 187–198.

Epting, S. (2017). On moral prioritization in environmental ethics: Weak anthropocentrism for the city. *Environmental Ethics*, 39(2), 131–146.

Epting, S. (2010). Questioning technology's role in environmental ethics: Weak anthropocentrism revisited. *Interdisciplinary Environmental Review*, 11 (1), 18–26.

Epting, S. (2021). *The morality of urban mobility: Technology and philosophy of the city*. London: Rowman & Littlefield International.

Flora, C. (2011). Review: Schanbacher, William D.: The politics of food: The global conflict between food security and food sovereignty. *Journal of Agriculture and EnvironmentEthics*,24, 545–547.

Guha, R. (1989). Radical American environmentalism and wilderness preservation: A third world critique. *Environmental Ethics*, 11(1), 71–83.

Guthman, J. (2011). *Weighing in: Obesity, food justice, and the limits of capitalism*. Oakland, CA: University of California Press.

Hampton, G.J. (2015). *Imagining slaves and robots in literature, film, and popular culture: Reinventing yesterday's slave with tomorrow's robot*. Lanham, MD: Lexington Books.

Hargrove, E. (1992). Weak anthropocentric intrinsic value. *The Monist*, 75(2), 183–207.

Jonas H. (1984). *The imperative of responsibility: In search of an ethics for the technological age*. Chicago: Chicago University Press.

La Via Campesina. (2007). *Declaration of Nyéléni*. https://viacampesina.org/en/declaration-of- nyi/.

Kemmerer, L. (2015). *Eating earth: Environmental ethics and dietary choice*. New York: Oxford University Press.

Lentini, N. (2022). Cancel culture: Trouble for brands or just noise? *Marketing Dive*, January 11. https://www.marketingdive.com/news/cancel-culture-trouble-for-brands-or-just- noise/616983/.

Mills, C. (2001). "Black trash." In L. Westra & B. Lawson (Eds.), *Faces of environmental racism: Confronting issues of global justice* (pp. ix–xxvi). Lanham, MD: Rowman & Littlefield.

Shiva, V. (1993). *Monocultures of the mind: Perspectives on biodiversity and biotechnology*. Penang: Zed Books.

Tremblay, R. (2010). *The code for global ethics: Ten humanist principles*. Amherst, NY: Prometheus Books.

Whyte, K. (2016). Indigenous food sovereignty, renewal and U.S. settler colonialism. In M. Rawlinson and C. Ward (Eds.), *The Routledge handbook of food ethics*. New York: Routledge.

7 The Intrinsic Value of Food Chains

Introduction

The previous chapters identified and unpacked myriad moral issues associated with food chains, making cases along the way holding why they should be of interest for reforms. To recall, actions centered on such efforts need to start with those who have already been harmed or remain vulnerable. A lesson inspired by mereology shows that we add, replace, or change parts of food chains that play significant roles in causing such harm. Focusing on part-to-part relations as they affect the entirety of a food chain provides a starting ground for conversations about how such parts cause numerous kinds of problems, which can impact several stakeholder groups.

The intense study of such arrangements exhibits how agribusiness contributes to wicked problems like climate change and vast social and economic inequalities. Within the context of wicked problems, labeling the troubling arrangements as "wicked technologies" helps us see how they exacerbate existing troubles. To remedy this situation, we need beneficial parts that fit the description of "saving" as often as possible. These technologies work towards or count as mitigatory efforts to deal with the aforementioned harmful outcomes.

Although there could be little evidence that these technologies make a difference, their mere existence suggests that the kind of thinking—saving thinking—a well-ordered enterprise aiming to alleviate harms stemming from wicked technologies, is underway, at least conceptually. Despite the inherent motivation embedded in this view, obvious criticisms will come. Of course, those that positively address ways to iron out conceptual wrinkles are welcome. However, one could make a case that the unproven track record of such enterprises is too significant to dismiss, meaning that, despite the harms of food chains, it is better to stick with a way that works. Delivering food across the globe is too important of an objective to risk an error coming from implementing a new standard of procedures.

We should not discount these reasons. However, criticizing unproven efforts for their embryonic situation is hardly fair, considering that any event in its beginning stages is where it should precisely be, which has made

DOI: 10.4324/9781003255505-7

nary an impact. Such comments show a lack of prepared thought and should be dismissed accordingly. They are little more than attempts to preserve the agricultural status quo that comes with the unethical baggage examined in Chapter 2. Further, embracing this attitude says that such harms should hold steady as collateral damage in the battle to make a buck while people continue to suffer, the nonhuman environment remains imperiled, and we risk the ability of future generations of people to live and thrive.

Responses like this attempt to misdirect the attention required to reform agribusiness rather than point out a theoretical inconsistency or practical issue that should override the need to change in the manner described above. While it is hopefully doubtful that a scholar would not attempt to make such weak arguments, it might serve us well to anticipate such views from some lawmakers, politicians, industry folk, and others who gain monetarily from business as usual. However, being aware that shills are likely to advance such arguments in the public sphere to preserve the status quo could help shape public opinion and encourage skilled professionals to undertake projects that support reforms.

Taking this point seriously means ignoring weak positions and working towards saving initiatives. Instead of accepting a litany of harm towards several stakeholders, the thinkers, engineers, farmers, logisticians, shippers, and architects engaging in the thinking that supports "saving endeavors" are unknowingly working towards agricultural redemption. These actions, in concert, count as humble steps illustrating how we start at the foundations, thinking our ways out of our wicked gastronomical existence. Moving forward, bolstering their efforts requires at least a two-pronged attack for the battle for an ethical future for food chains. Despite the optimism embedded in the tone above, ushering in such a bold restructuring must be executed keenly yet judiciously.

Engaging in such practices should be done with great care because human lives are at stake in these affairs. Hasty decisions that deal with food security could be deadly, but this reality should not deter reform efforts. In turn, this chapter's purpose is to illustrate these points, modeling two ideas for agricultural technologies that could help advance needed reforms. It examines an overlooked motivating element that underpins reforms: intrinsic value. Next, it explores two kinds of saving agriculture technologies that differ drastically, showing how they could bolster the measures that could help reshape agribusiness according to moral ordering.

The first is a physical part, vertical farms. Early models exist, and preliminary evidence suggests that their further development and implementation could yield outcomes that eliminate or reduce harm as they affect stakeholder groups. The second is nonmaterial, but it has the power to improve the conditions associated with food chains at their first links, all the way to the utensils. This "sustainable food" label is essentially a policy represented by a label. One backed by governmental policy, regulation, and enforcement, could revolutionize food chains, possibly changing the harmful, unethical conditions associated with them. While similar-minded labels have met with significant criticism and arguably failure, the goal here is to farm some freestyle thought, a kind of thought experiment. The goal is to see the drastic efforts required to reform food chains suffering from innumerable immoral failures. This chapter connects and fleshes out these points, showing that another food world is possible, even though it is incredibly far from today's agribusiness. To help facilitate such bold ambitions, I explore some future efforts to support these endeavors.

The Intrinsic Value of Food Chains

While the issues related to food chains examined in Chapter 2 were upsetting, saddening, and gloomy, the topic explored here shows powerful reasons for hope. Despite the ill conditions found throughout food production and distribution, those aspects are not permanent. They are merely temporary elements that have received support. The associated parts that cause harm and perpetuate suffering are replaceable, meaning that saving technologies can replace them or help with the needed efforts. Although this line of thought aims to secure practical mitigatory steps, an underlying dimension reveals aspects of food chains that matter—beyond the material resources that agribusiness provides. Namely, I am talking about the intrinsic value of food chains. I think about intrinsic value as it aligns with how it was hashed out in the environmental ethics literature while that subfield was taking shape.

For instance, Eugene Hargrove (1992) argues that there are two kinds of it, "objective" and "subjective." The former entails that intrinsic value exists independent of humankind's recognition of this value, and the latter depends on humans assigning it. While there is much appeal to objective intrinsic value, proving that such values exist is riddled with challenges. One could argue that this type of intrinsic value does not provide an understanding of intrinsic value required for grasping why it matters for food chains. Engaging in this line of thought actually impedes conversations necessary for establishing criteria for what deserves protection and preservation. These discussions could bog down the relevant topics of interest, focusing on *how* we can know that such value exists, which could be an unending ordeal. However, subjective intrinsic value provides a way for humans to determine what has and lacks intrinsic value. In dealing with the nonhuman world, Hargrove holds that ecological experts are qualified to make such assessments.

Here is an example that aligns with this thinking. Consider the original US Constitution.[1] For US citizens, it has tremendous instrumental value in that it dictates the nation's highest laws, illustrating the operational procedures of the government and its limits, and protects people's rights and freedoms. The original document is widely accessible in print and on the internet. The actual first documents are stored and protected. Experts could argue that new copies have no intrinsic value, but the same argument would not hold for the original even though we have secured its instrumental value through having numerous copies. Experts (and almost everyone else, probably) would argue that we must preserve the original document for its intrinsic value. It is valuable aside from any instrumental value, which other, newer copies ensure. When it comes to experts, their views are incredibly significant, especially when dealing with historical and culturally substantial elements (i.e., authentic versus counterfeit).

The pattern behind Hargrove's conception of intrinsic value works well when applied to examining food chains. Those who understand their complexity and know those aspects intimately also have a kind of expertise that gives their views credibility and buy-in that gives them a vested interest in determining subjective intrinsic value of the patterns underlying food chains. Understanding how to conceptualize the entailments of a food chain is beneficial for seeing this value fully. However, the worry is that if we only define a food chain technically, a textbook account will list the parts and operations, which might provide a more detailed understanding of what we expect to find associated with such conceptions. However, going beyond this narrow view reveals more aspects that we need to study. For instance, philosophically interrogating the intrinsic value of food chains reveals elements that are of paramount importance for understanding why we should care about them in the first place. The process also shows why we should fight for them and make food chains that we should celebrate because they serve humanity.

Understanding this point requires that we examine a tension. For example, Chapter 2 shows that people, natural resources, and numerous physical and nonmaterial technologies are necessary to produce and distribute food worldwide. People and nonhumans are harmed in the process. What is understood but also understated is that such processes represent the apex of humankind's collective intellectual enterprises working in concert to provide the means to survive and thrive. I am not talking about the ability to oppress others. That is horrible. I am referring to the capacity to grow and distribute food in such fast-moving and sophisticated ways. Not just nutrition but survival in this context includes the numerous ways we benefit from food, from earning a living to how people connect with and maintain cultures, sometimes across vast distances. The tension here is that while food chains help secure many good ends, sometimes the employed means are not. In some cases, they are despicable.

The good parts that help us grow and distribute food in impressive ways require looking at them as they are found in food chains. We see that they exist most fully with the optimal applications of mathematics, logistics, engineering, sciences, policy, the internet of things (IoT), and business. They form a confluence that makes the modern world possible regarding food security. In addition, they also rely on the labor of 28 percent of the Earth's total population, about a billion people (Cassidy and Snyder 2019). Despite the instrumental value that the above fields deliver, we can assign intrinsic value to food chains—after removing harm. In this light, they are good in and of themselves. Let us continue to think about food chains as if they do not cause harm. By employing technologies such as advanced mathematics, engineering principles, and cultural histories, food chains combine these elements to make seemingly impossible gastronomical adventures realities that we can enjoy without much thought or appreciation. Food chains make such conditions and circumstances achievable, even though otherwise they would not exist.

At a macroscale, the ability to orchestrate countless shipments of food that crisscross the entire globe *innumerable times daily*. Countries that lacked food security could achieve it through food chains (D'odorico et al. 2014). Yet, people tend to experience the benefits of food chains personally. I am talking about immigrants' ability to enjoy culturally significant foods far away from their homelands. The actuality is that people can share their loved culture with others or make a living doing it. There is also the ability of people to enjoy foods from places they long to visit and for foods to fuse in ways that deliver tantalizing experiences. They can enjoy fresh fruits and vegetables on demand or out of season in frozen lands or while orbiting or leaving Earth. Foods now can travel extreme distances without spoiling due to various methods. Children living in arctic climates can receive nutritious foods consistently.

While people enjoying these benefits could agree that they like having these foods outside of familiar contexts, they could also hold that the ability to have them is good in and of itself. After thinking about food chains with the bad parts replaced with saving ones, the goal for agribusiness reform should be clear. It is to *secure* and preserve the intrinsic value of food chains while removing the immoral parts and parts that cause harm and contribute to wicked problems such as climate change and vast social and economic inequalities. By drawing attention to "securing," the point is that most food chains at present may have intrinsically valuable properties. However, at the same time, we cannot claim that they are intrinsically valuable without also holding that harming people and nonhumans is intrinsically valuable. These points are irreconcilable. Removing or replacing bad parts must happen for food chains to have intrinsic value. Such processes must hold steady as perpetual enterprises.

In Chapter 1, I mentioned that justice in this context is not something we can achieve, but it is a continual process worth pursuing, which is valuable in and of itself. On the surface, this point seems inconsistent with the idea that food chains have intrinsic value if we can remove their harm. Food justice is an entirely doomed undertaking if that situation holds indefinitely. However, this view is misguided. Suppose we consider pursuing justice as an intrinsically good activity. Then, we think about it as a dimension of food chains that we attend to by including moral ordering when re/designing and *maintaining* food chains. The justice aspect in "maintaining" focuses on the role of justice applied to food chains. This usage entails that we think of it as a verb. Justice is an action rather than an achievable milestone. Including this caveat provides a way for food chains that would otherwise count as being good, ethical, or just that can falter on occasion. Nothing is perfect all the time.

Mistakes can happen that involve harm is acceptable, albeit unwanted. Holding that humans cannot make them puts unreasonable and unrealistic expectations on food chains. Yet—by adjusting a working conception of them that brings this element into their conceptual structure, we morally improve their status. In turn, food chains are sophisticated systems, and mistakes are bound to happen. What matters here is that mitigation and amends are integral as well. This point shows how food chains can have an intrinsic value worth celebrating.

We should not need food chains' intrinsic value to motivate us to remove the bad parts. It is doubtful that all food-justice activists will care about this notion when stacking it next to the actualities of harm. Still, it counts as one modest contribution towards seeing the panorama of food chains' total values. Regarding harm, though, accounting for food chains' intrinsic value also reveals another aspect of harm and wrongdoing. Taking inventory of agribusiness' kinds of harms should also include eliminating or at least challenging food chain's intrinsic value. If this value is not established and recognized because of the immoral acts happening along food chains, then food chains are robbed of such values due to these unfortunate realities. We can add this charge to the list of wrongs in Chapter 2. Either that or the integrity of such values is inherently degraded.

However, we can further flesh out this view to reveal additional insights into the significance of food chains' intrinsic value. The following section does just that. It focuses on the ontological nature of food chains' intrinsic value, exhibiting that it is remarkable—or at least it could be with reforms. In turn, it gives us another reason and motivation to work towards removing bad parts and celebrating food chains that illustrate how to maximize humankind's ability to engage in processes that highlight how technology and morality can make progress in unison.

Super Intrinsic Value?

Although it might seem like examining food chains' intrinsic value provides an encompassing view of their significance, there is another dimension that we can reveal by examining the intrinsic value of food chains. One could argue that numerous aspects mentioned above (e.g., mathematics) have their own intrinsic value. They are valuable beyond their instrumental values. The mathematics that helps with logistics is intrinsically valuable because they help us make needed calculations, but mathematics is simply good aside from this value. Kids are taught this lesson in classrooms when they ask the teacher if they will ever use this math.

The pattern of this argument also applies to other bodies of knowledge employed in food chains. With this point in mind, it would not be far-fetched to hold that food chains—as a confluence of apexes regarding humankind's mental and physical technologies—have super-intrinsic value. The combined intrinsic values are more than an isolated appreciation of a single intrinsic value. Food chains have their own intrinsic value, but what makes them unique is this confluence of intrinsic values that make it possible for them to exist. Due to this extraordinary value, food chains deserve not just celebration for themselves but also increased attention and study for optimization.

Of course, this argument is not something we can "prove," but aiming to prove it misses the point. The notion that needs spotlighting is that we are trying to uncover the depth of why food chains are valuable beyond food production and distribution. They are good for their own sake. Saying that they have super-intrinsic value is not claiming that some mysterious feature emerges. Nothing is in the whole that is not in the parts. This statement merely indicates the kind of things that we are dealing with when thoroughly examining the ontological character of a food chain's intrinsic value. Nothing more.

The problem is that, at present, food chains carry the associated harms as examined in Chapter 2. The goal, then, should be to replace the bad parts with ones that deliver more ethical expressions of humanity's best thinking regarding the areas noted above. As mentioned, this process starts with technologies of the mind, the ontological and value systems that we employ to assess the quality, integrity, and morality of food chains. In those regards, the goal is to, part-by-part and case-by-case, create arrangements within food chains that significantly reduce or eliminate harms as investigated. Moral ordering can guide the process, serving as the technology that can help deliver the desired outcomes that reflect the kind of people we want to become. We should want food chains that represent the best thinking in relevant areas while also delivering food in ways that do not involve immoral or harmful acts.

One step in the right direction is identifying parts that can help secure such outcomes. The following section provides two examples, vertical farming operations and sustainable food labels, which will help unpack the notions above. The former has early models that we can examine, but the latter is imaginary. However, when we study the evolution of food labels, the next logical advancement is close. Exploring these technologies illustrates how they can help deliver better outcomes and provide ways to make ethical choices.

Vertical Farming

The "food miles" problem is a concern for any food-chain-related operation. Such issues require attention on a case-by-case, or part-to-part, basis. For instance, in some cases, importing a fruit or vegetable would support moral ordering more than growing it locally. Dealing with logistics and resource allocations is beyond most philosophers' abilities, meaning that only trained experts should weigh in on those matters. However, examining near examples that exhibit better outcomes that could address the issues found in Chapter 2 is an activity open to anyone. Vertical farming is an exemplar, which aligns well with a description of a saving technology in many cases. Although such devices are relatively new, some offer promise, and others are solid based on their results.

Defined loosely, vertical farming means any growing operation that goes up rather than horizontally. The design inherently seeks to maximize space and control the elements vital to food production, such as nutrients, light, water, and pests. On the smaller end of the scale, individual farming towers can serve niche grocery stores, restaurants, and local entrepreneurs. Citizen groups, municipalities, and ambitious individuals wanting to shape their food security could also benefit from this technology, lessening demand from agribusiness (Epting 2016). These devices already serve an educational purpose in some places, teaching children about technological approaches to food production in ways that challenge conventional means (Fernandez 2016).

One of the more successful, large-scale operations is Sky Greens in Singapore. This operation's 22 vertical towers yield five times more food than an equal horizontal space could provide, using a mere one kilowatt of electricity each hour (Krishnamurthy 2014). Since then, numerous enterprises worldwide have followed Sky Greens' success. There are enough examples to study that academics have reviewed them as a category, showing how they can support sustainability, especially in urban environments (Kalantari et al. 2018).[2] Recently, retailers have invested significantly in these growing operations, hoping to bolster their supply chains (Repko 2022). There will be numerous grounds to critique such moves. Luckily, academics thrive in this area. Nevertheless, if these measures can significantly replace bad parts, exploring options to promote them deserve attention and study. As stated previously, each case requires assessment, focusing on moral ordering to ensure that stakeholder groups receive benefits and harm reductions proportionate to their suffering and condition within the moral order.

Still, there are significant reasons to champion these devices due to the numerous positive aspects inherent to the design or that emerge due to socio-material and environmental situations. For example, the increased ability to eliminate pesticides, improve working conditions, expand affordability for consumers, lessen harmful impacts on ecosystems, reduce pollution, costs, energy demands, and secure more food, vertical farming operations large and small have promise. Granted, such operations are limited to specific kinds of foods such as lettuce, meaning that growing corn might be seriously limited.[3]

However, keeping in mind the nascent state of this technology, it is not far-fetched to say that advancements later will be forthcoming. If some or all of the above improvements materialize, then maintaining these devices qualify as saving technologies remains sound. In turn, there is significant motivation to support their implementation as new parts that can replace troublesome ones—or as new parts joining food chains as powerful missing links that can deliver benefits in several ways to many stakeholder groups.

It seems that businesses have learned about financial incentives for supporting vertical farming as "saving enterprises." These technologies count as only one possible way to reform agribusiness ethically. As similar devices enter the marketplace, the effects could create a better world wherein stakeholder groups are not treated poorly. In turn, identifying measures that speak to this point deserve encouragement across scales. Some such measures might not exist, but existing devices could point in that direction, with the following steps being to look for advancements that can have more substantial impacts. One such imaginary technology is a sustainable food label that could carry the power of regulation and enforcement. The section below provides a preliminary investigation into the possibility that this kind of saving technology could bolster reform efforts to benefit all identified stakeholder groups.

Food Labels as Agricultural Technologies: Towards a "Sustainable" Food Label

It might seem odd to think about a food label as a technology but doing so makes sense. A label should not seem weird at this point, having already dealt with well-ordered thinking as a kind of technology. If we stick with the shared conception of the term in a broad sense, technologies are things that we invent to help us complete a job. One way to view food labels is to see them "help" us decide which foods to eat based on dietary preferences. Over the last few decades, global markets have witnessed numerous labels don the fronts, backs, and sides of food and food-like products. They are worldwide.

For instance, along with nutrition facts, there are also warnings in some countries, such as Mexico, indicating that an item has excessive sodium, sugar, or fat (Esposito 2020). Some nations warn about genetically engineered foods or genetically modified organisms (or GMOs) (Center for Food Safety 2022). One label mentions being non-GMO (The Non-GMO Project 2016). Other labels inform us that the foods are organic (United States Department of Agriculture 2021). Eggs now come with several labels, covering everything from housing and activities to breeding encounters. They include but are not limited to the following: "United Egg Producers Certified," "United Egg Producers Certified Cage-Free," "American Humane Certified," "Certified Humane," "USDA Certified Organic," "Cage-Free," "Free-Range," "Pasture-Raised," "USDA Grade," "Antibiotic-Free,"

"Pasteurized," "Enriched with Omega-3," "Vegetarian Fed," and "Fertile" (United Egg Producers 2022).

Fish is labeled as farmed or wild-caught, and some tuna labels indicate if fishing poles captured them (Leschin-Hoar 2017). Other fish labels appear to rally against overfishing (a harmful practice). They allege to be "sustainable" (but their use of the term leaves much to be desired for some) (Marine Stewardship Council n.d.).[45] Labels on packages of cow meat let you know if the animal ate grass (United States Department of Agriculture 2019). Select supermarkets have unique systems proclaiming the humaneness of the meat for purchase. For example, one grocery store chain in the US advertises several levels of humaneness, and you can purchase cow meat made from a cow that spent her entire life on the same farm (Whole Foods Market 2022).[6] They let customers know that they deserve to have positive feelings when buying their meat. Some products claim to support rainforests sustainably (Rainforest Alliance 2020). For most of these labels, there is someone, an organization, or a whistleblower ready to cry foul or criticize the definition or metric that the label makers use (e.g., Food Integrity Campaign 2022).[7]

Considering the above points in tandem, we can say that food labels can be a promising technology that works well. People wanting to lower their sugar intake can quickly gauge their consumption with a glance, maximizing their dietary agency. If consumers want to think that their eggs came from happy chickens, a carton made from recycled materials can help them feel good about their decision. Conversely, if a brand wants to mislead customers to cash in on a trend such as organic, they can do so. For instance, by adding the phrase "made with organic ingredients," which only requires some of many ingredients to be organic, one could say that such labels are "bad" technologies because they help nefarious biscuit makers deceive unsuspecting biscuit eaters.[8]

Considered in the narrow context of customers in grocery store aisles, perhaps a food label on an individual product can help consumers make choices that partly embody their values or help them with dietary agendas. While on this scale, the effects would be marginal. However, the impacts could add up across a nation with millions of people. Again, the outcomes could increase by putting the same label on similar items. This example is rhetorical, but the "could" is why it is given. This point signals what is at stake. At present, if people want to support sustainable agribusiness, they lack a feasible ability to know what products deserve the adjective.

Further, most customers cannot define it arguably. Yet, these conditions do not entail that reform efforts should not strive for it, especially considering the ethical outcomes that depend on it. Still, the possibility could or should exist that allows customers to shop in a manner that supports extensive ethical choices that impact the moral order of stakeholders per the earlier chapters while encouraging others to follow suit.

At present, customers wanting to buy items that align with such measures would have to look for products with several labels on a package, and many such labels could be empty when examined closely. As a mere part of a food chain, the ability to help in the decision-making process could have significant impacts if ushered in carefully, mindfully, and honestly. A sustainable food label could do just that. It would serve as a part that could work backward while looking forward. In a much broader, geopolitical context, a sustainable food label would require enforcement in the fields.

To adorn the sustainable mark, workers at each link in a food chain would have to weigh in on their wages, working conditions, and benefits approvingly. Consumers would know that a food or food-like product is not inherently unhealthy. Environmental impacts would be fitting, according to ecologists and specialists who have concerns for the biosphere and its future people firmly in view. Land, real estate, neighborhoods, and urban infrastructures and structures could not be tainted with ill-treatment, and residents must be able to participate in such decisions meaningfully. These criteria are extensive and require expertise going beyond a specific discipline.

Recall the expansive nature of agribusiness issues intersecting with climate change, a wicked problem. In turn, the requisite knowledge to label a product as "sustainable" calls on several veins of knowledge, bringing in numerous players from across the academy, industry, and government. The necessary undertakings will require advanced study to the point wherein even preparing for such a venture will require extensive preliminary investigation to determine the tentative parameters needed to develop a guide to discover the required knowledge and how it fits together. As a humble beginning, the following section moves in that direction.

Future Areas of Required Research

As a philosopher, touching on the required specializations needed to assess what counts as sustainable beyond ontological, social, or ethical dimensions would be out of line. Still, it is reasonable to note that philosophers and justice researchers could contribute to how a sustainable food label materializes and what it would entail. To see such benefits, consider applying the process of moral ordering to food labels and what it would mean for a sustainable food label. Recall how moral ordering works. Actions should be prioritized in the following manner: First, think about marginalized, harmed, or vulnerable people. Then, consider the public, nonhumans, future generations, and artifacts. It is nice to see food labels because they reflect thoughtfulness and care, but, seen collectively, the totality of food labels (if we could lump them together) reveals a kind of problem of moral prioritization. For instance, seeing them as a "continuous panorama" could show that, at present, food labels favor nonhuman animals.

For instance, one way to think about food labels is as preemptive responses to customer-based inquiries: Regarding chickens, they might ask:

"Were these chickens kept in cages?" Or, "Could they roam on the range?" For livestock, the questions could be, "Was this cow raised in a pasture?" "What did this cow eat?" "What about her diet?" In terms of pigs, "Did this pig live in a crate?" The labels that correlate to these questions suggest that people are willing to pay more to know that their animals lived "quality" lives, had good diets, could move freely and were free from cages.

There are no food labels that say something like, "farmworkers receive respectable wages and health insurance," "employees are paid well and have stock options," or "meat processors receive respectable wages, health benefits, and retirement saving programs." These labels do not exist, which could suggest that no correlative questions are lurking in customers' minds (at least not detected by research marketing groups).

Moral ordering, applied to the development of a sustainable food label backed with legal enforcement measures from an entity such as the USDA, could prioritize the ethical treatment of farmworkers *worldwide*. This point entails that it does not matter where a person works. They could work in Missouri or India. If bosses exploit workers or coerce them to work in horrible environments, the product is not sustainable. The social dimension of sustainability is not met. Approaching this step with respect and integrity means that the people who work tirelessly in fields, factories, transportation centers, distribution hubs, and warehouses can maintain a living, see doctors, dentists, optometrists, and afford to take a family vacation when fitting. Such measures should be the first step towards securing the conditions for applying a sustainable food label. While making this assertion seems straightforward, it entails monumental efforts to ensure meaningful outcomes that measure up. Determining metrics, accountability, and transparency will hold steady as challenging undertakings, especially internationally. Future work should focus on finding arrangements of parts (i.e., food labels) that support better (more ethical) outcomes.

The next motivation should be to determine if the foods are healthy, relying on nutritionists, public health experts, and medical professionals. Suppose a sustainable label is on a product. This label would let customers know that relevant experts had thoroughly vetted this food. In that case, the customer will understand that, if eaten in healthy amounts, this food will help them meet their daily nutrition goals while not subjecting them to unhealthy amounts of sugar, sodium, or unhealthy fats.

Determining standards for acceptable impacts on nonhuman nature is categorically outside most philosophers' skillsets. Such measures will require expertise from relevant fields. Ecologists, biologists, and other highly specialized researchers will need to weigh in on agribusiness's effects on the nonhuman world, employing their insights to determine acceptable and prohibited actions to secure the sustainable label. Most of the aspects that involve this stakeholder group fall outside philosophers' purview. However, contributions from relevant researchers such as Hargrove (1992), who deals with intrinsic value, could help establish criteria for how and when to appeal to the intrinsic value of nonhuman life when conflicting situations arise. The benefit of this measure is to properly secure a place for the nonhuman world within the process of moral ordering that does not give preference for nonhuman life above human life, except for exceptional cases that require it. This point reminds us that moral ordering is suggestive and not absolute, meaning that it can adjust to deal with the world's conditions rather than forcing the world to conform to our expectations of it, which becomes more complicated when trying to feed human populations.

While there are good reasons to provide moral consideration for the previous stakeholder groups, it is challenging to say that future generations "deserve" it. The worry is that they do not exist, and it is complicated to say that we owe something to people who do not live. However, Hans Jonas (1984) developed a conceptual workaround, holding that we (basically) have a duty connected to his earlier imperative, which includes preserving the idea of humankind existing in the future. In this way, it makes sense to say that we owe it to ourselves to maintain the environmental and social conditions for their existence. When examining the initial definition of sustainability as worded by the World Commission on Environment and Development (1987, 41), we notice that moral ordering aligns with this description. "Sustainable development is development that meets the needs of the present without compromising the ability of future generations to meet their own needs."

Although this description is open to interpretation and numerous ways when putting it into policy, it supports the idea that a sustainable food label should adhere to moral ordering, putting future people in this position. The pattern behind the language substantiates the strategic ordering when unpacked. For instance, future generations come after those existing now, meaning that the moral ordering behind a sustainable food label, as laid out above, works. The positioning of future generations after existing groups of humans and the nonhuman world will only exist if the ecological and social conditions make future generations (hopeful) existence possible.[9] In turn, this stakeholder group ends up in its place in the moral order by default.

Lastly, and by default, humankind's artifacts come last in the process of moral ordering for consideration of a sustainable food label. This reason is that, quite simply, they are replaceable in most instances. However, cultural and historical artifacts of significance require examining their place in the moral order. Such dimensions demand further exploration and conversation with the people to whom these artifacts hold importance in their lives, including intrinsic values and instrumental ones. This point forces us to recall moral ordering's suggestive nature. This quality makes it possible to have moral insights while maintaining a degree of flexibility that avoids having a process designed for a particular purpose that pushes against it. Asked rhetorically, what is the point of developing moral guides for action if the results of such actions go against the initial motivation for the guide? The point is for the guide to serve us instead of serving it.

Having fleshed out the moral ordering behind implementing a sustainable food label counts as humble steps towards its conceptualization. Still, much of the "heavy lifting" remains. Most of this work falls outside of philosophy's boundaries, suggesting the interdisciplinary nature of such a project. However, a philosophical examination could benefit this work by accompanying it along the way. The point here is not to interfere, but providing ontological, social, and ethical insights could prevent future missteps and proactively guide the work. Considering that the parts of food chains come and go, as is their inherent character, examining them in ethical manners must keep pace. This notion implies that the philosopher of food's work must endure, being just as resilient as the food chains themselves.

Conclusion

This chapter brought the key notions developed in this text together, illustrating an avenue for ethical agribusiness reform. It showed how examining part-to-part relationships, and planetary conditions such as wicked problems broke those issues down to the unethical outcomes affecting several stakeholder groups. Rather than simply complain about such situations, the work here begins to deal with these matters through developing ways to conceptualize the problems that also gestures towards mitigatory steps, wicked and saving technologies.

Two examples of possible saving technologies were provided to exhibit their vast differences: vertical farming and a government-backed sustainable food label. These devices emblematize the kind of thinking required for ethical reform, which begins with the sort of well-ordered thought that can guide the needed steps to deliver us from an anthropogenic demise. During this process, I made the case that food chains are not simply instrumentally valuable, but they can be intrinsically valuable if harms are removed.

However, because several other entities bearing intrinsic value work in concert to create food chains, referring to them as having super-intrinsic value makes sense. It is a way to describe them that accounts for the ontological conditions of food chains' intrinsic value. As an entity that bears this intrinsic value, it deserves protection. Yet, examining how it manifests at present also shows that we have allowed food chains to exist that produce harmful outcomes, which is wrong. Yet, thankfully, such situations can be temporary. Employing the process of moral ordering as applied to the implementation of saving technologies can help us rid food chains of these unwanted harms and create food chains that show respect and moral consideration for the groups of stakeholders they affect.

Notes

1 I used this example elsewhere. See Epting (2019).

2 It is worth pointing out that Kalantari et al. (2018), while employing the term "sustainability" in the review process, neglect to focus significantly on elements of vital importance to sustainability such as environmental justice, which scholars argue most hold steady within such conceptions. Failing to do so could be problematic if this aspect is overlooked. For more information, see Agyeman et al. (2002).

3 I am grateful for a student who mentioned this point about corn in my "Creating Future Cities" class, a philosophy course that I created at the Missouri University of Science and Technology. Due to privacy regulations, I cannot mention his name. Still, I appreciate his insight and want to acknowledge it in some capacity.

4 While it does appear that Marine Stewardship Council applies rigorous standards when it comes to science, they admit that the socio-economic elements receive less study. Due to this aspect, one could seriously challenge their use of the term "sustainable," and one could go as far to say that their use of the term leans towards a similar term to greenwashing. For instance, Marine Stewardship Council (n.d., para. 2) mentions: "The impact of the MSC program on environmental sustainability is an established area of research. However, the social and economic effects that occur as a result of certification are less well understood." For more information, see Marine Stewardship Council (n.d. 2).

5 It is worth mentioning that seafood labeling is rather troublesome. Activist groups, for instance, have called for regulations such as the one that I argue for in this chapter. For example, Food & Water Watch (2010, para. 7) indicate: "The eco-label certification programs reviewed in this report demonstrate inadequacies with regard to some or all of the following: environmental standards, social responsibility and community relations, labor regulations, international law, and/or transparency."

6 It is worth mentioning that one could argue that meat is not or never be sustainable due environmental and or normative considerations. Again, as stated in Chapter 6, this conversation would require substantial study and debate.

7 One could argue that the closest label to a "sustainable" one is "Fair Trade," which *seems* like a substantial step in the right direction towards reform (Fairtrade International n.d.). However, researchers suggest that this label has *several* challenges, and there are many issues to examine (Niemi 2010; Carimentrand and Ballet 2010; Valkila et al. 2010). Further, Fair Trade is basically a research strand with numerous areas of contention and debate in academia and beyond. For more information on the history of Fair Trade, see Dragusanu et al. (2014).

8 "Bad" in this context only means within specific socio-material situations.

9 The pattern of thinking behind the 1987 Bruntland Commission Report, authored by World Commission on Environment and Development (1987, 41), could also support that many of the world's farmworkers, being poor and harmed in numerous cases, require that moral ordering favor their positions in food chains, meaning that agribusiness reform efforts should give then initial consideration and action. The report holds: "Sustainable development is development that meets the needs of the present without compromising the ability of future generations to meet their own needs. It contains within it two key concepts: the concept of 'needs', in particular the essential needs of the world's poor, to which overriding priority should be given[.]"

References

Agyeman, J., Bullard, R., & Evans, B. (2002). Exploring the nexus: Bringing together sustainability, environmental justice and equity. *Space and Polity*, 6(1): 77–90.

Carimentrand, A., & Ballet, J. (2010). When Fair Trade increases unfairness: The case of quinoa from Bolivia. *Cahier FREE*, 5, 17.

Cassidy, E, & Snyder, A. (2019). Map of the month: How many people work in agriculture? Resource Watch. May 30. https://blog.resourcewatch.org/2019/05/30/map-of-the-month- how-many-people-work-in-agriculture/.

Center for Food Safety. (2022). International Labeling Laws. Center for Food Safety. https://www.centerforfoodsafety.org/issues/976/ge-food-labeling/international-labeling- laws.

D'odorico, P., Carr, J.A., Laio, F., Ridolfi, L., & Vandoni, S. (2014). Feeding humanity through global food trade. *Earth's Future*, 2(9), 458–469.

Dragusanu, R., Giovannucci, D., & Nunn, N. (2014). The economics of fair trade. *Journal of economic perspectives*, 28(3), 217–236.

Epting, S. (2016). Participatory budgeting and vertical agriculture: A thought experiment in food system reform. *Journal of Agricultural and Environmental Ethics*, 29(5), 737–748.

Epting, S. (2019). Automated vehicles and transportation justice. *Philosophy & Technology*, 32(3), 389–403.

Esposito, A. (2020). Mexico's new warning labels on junk food meet supersized opposition from US, EU Reuters. August 11. https://www.reuters.com/article/us-mexico-health- idUSKCN25802B.

Fairtrade International. (n.d.). What is Fairtrade? Fairtrade International. https://www.fairtrade.net/about/what-is-fairtrade.

Food Integrity Campaign. (2022). Transparency. https://foodwhistleblower.org/issue/transparency/.

Food & Water Watch. (2010). De-coding seafood eco-labels: Why we need public standards. Food & Water Watch. https://www.ftc.gov/sites/default/files/documents/public_comments/guides-use- environmental-marketing-claims-project-no.p954501-00152%C2%A0/00152-56693.pdf.

Fernandez, D. (2016). Oak Cliff Students use vertical farming lesson to help community. WFAA 8 ABC, April 21. https://www.wfaa.com/article/news/education/oak-cliff- students-use-vertical-farming-lesson-to-help-community/287-148022886.

Hargrove, E. (1992). Weak anthropocentric intrinsic value. *The Monist*, 75(2), 183–207.

Jonas H. (1984). *The imperative of responsibility: In search of an ethics for the technological age.* Chicago: Chicago University Press.

Kalantari, F., Tahir, O.M., Joni, R.A., & Fatemi, E. (2018). Opportunities and challenges in sustainability of vertical farming: A review. *Journal of Landscape Ecology*, 11(1), 35–60.

Krishnamurthy, R. (2014). Vertical farming: Singapore's solution to feed the local urban population. *Permaculture News*, July 25. Permaculture Research Institute. https://www.permaculturenews.org/2014/07/25/vertical-farming-singapores-solution-feed-local-urban-population/.

Leschin-Hoar, C. (2017). Will fish get a humanely harvested label? These brothers bet $40 million on it. *Food for Thought*. https://www.npr.org/sections/thesalt/2017/06/14/532845573/will-fish-get-a-humanely-harvested-label-these-brothers-bet-40-million-on-it.

Marine Stewardship Council (n.d. 1). What does the blue MSC label mean? Marine Stewardship Council. https://www.msc.org/what-we-are-doing/our-approach/what-does-the-blue-msc- label-mean.

Marine Stewardship Council (n.d. 2). Monitoring socioeconomic impacts. Marine Stewardship Council. https://www.msc.org/what-we-are-doing/science-and-research/monitoring- socioeconomic-impacts-of-our-program.

Niemi, N. (2010). "Empowering Coffee Traders? The Coffee Value Chain from Nicaraguan Fair Trade Farmers to Finnish Consumers." *Journal of Business Ethics*, 97(2): 257–270.

Rainforest Alliance. (2020). What Does "Rainforest Alliance Certified" Mean?https://www.rainforest-alliance.org/insights/what-does-rainforest-alliance-certified-mean/.

Repko, M. (2022). Walmart makes an investment in vertical farming start-up Plenty. *CNBC*January 25. https://www.cnbc.com/2022/01/25/walmart-makes-an-investment-in-vertical- farming-start-up-plenty.html.

The Non-GMO Project. (2016). About. The Non-GMO Project. https://www.nongmoproject.org/about/.

United Egg Producers. (2022). Store labels UEP certified guidelines. United Egg Producers. https://uepcertified.com/store-labels/.

United States Department of Agriculture. (2019). What is "grass fed" meat? United States Department of Agriculture. https://ask.usda.gov/s/article/What-is-grass-fed-meat.

United States Department of Agriculture. (2021). Understanding the USDA organic label. United States Department of Agriculture. https://www.usda.gov/media/blog/2016/07/22/understanding-usda-organic-label.

Valkila, J., Haaparanta, P., & Niemi, N. (2010). Empowering coffee traders? The coffee value chain from Nicaraguan fair trade farmers to Finnish consumers. *Journal of Business Ethics*, 97 (2), 257–270.

Whole Foods Market. (2022). Meat Department Quality Standards. https://www.wholefoodsmarket.com/quality-standards/meat-standards.

World Commission on Environment and Development. (1987). Report of the World Commission on Environment and Development: Our Common Future. https://sustainabledevelopment.un.org/content/documents/5987our-common-future.pdf.

8 Conclusion

Introduction

Although the preceding chapters examine agribusiness reforms thoroughly, a few remaining insights emerge from looking at the primary notions and terms quickly in tandem. These topics include a variety of agricultural harms found in agribusiness, applied mereological insights, wicked and saving technologies, moral ordering, food chains' intrinsic value, and super-intrinsic value. Looking at these notions panoramically shows some of the common characteristics one could expect to find when examining saving technologies in the fields, roads, tracks, hubs, seas, stores, and restaurants. The motivation behind this review is to encourage future research that can put these ideas to use, which can increase the ethical integrity of agribusiness. To that effect, this chapter's purpose is to review them.

Significant Themes and Key Terms: How They Fit Together for Moral Progress

Understanding the ethical dimensions of agribusiness requires knowing about the variety of issues that one finds when surveying the terrain. In turn, Chapter 2 provided an overview of the topics. It examined issues that affect workers at the beginning of food chains from fields to processing centers and beyond. The most disturbing problems include modern slavery, child labor, pesticide exposure, and worker safety. This chapter also looked at how agribusiness impacts consumers at the other end of the chain with harmful public health issues. While there are many harms to humans, the nonhuman world is not excluded from these worries. Topics such as fertilizer and chemical runoff, deforestation, species displacement, and climate change, in general, pose significant moral challenges. Issues in the nonhuman environment could also impact future generations, raising concerns for them. Lastly, tensions between cities and agribusiness illustrate that, as cities expand and compete for resources, ethical discussions about topics such as water allocation and land use will require attention.

While there is value in identifying ethical issues associated with food chains, efforts cannot end there. However, due to the overwhelming array of

DOI: 10.4324/9781003255505-8

cases, there is a need to deal with those matters effectively. I employ the thinking inspired by "applied mereology," an approach that can help us deal with part-to-part relations. This move breaks down how parts of food chains to see how problems emerge. Seeing food chains' operational structure through this lens shows that the problem of globalized opacity, having too many parts spread across vast distances, challenges our abilities to learn of morally complex arrangements of parts. While distance and number are not inherently troublesome, these situations easily lend themselves to such affairs. In turn, reform efforts should aim to increase transparency through various forms to increase accountability and pinpoint areas of concern.

Having a way to situate and deal with harm and food chains means gaining a way to make progress. Moving forward entails providing additional analysis. It needs to put agribusiness in perspective showing the scale of harm. The first step looks at wicked technologies that play a role in wicked problems such as climate change and vast social and economic inequalities. With the idea of these devices in focus, the attention turns toward saving technologies. Their design and subsequent outcomes can mitigate existing harms and promote goals consistent with sustainability. While these technologies show promise, the additional concern is that implementing them requires safeguarding against the problem of moral prioritization.

This issue means that when dealing with several stakeholders—such as farmworkers, consumers, nonhumans, and future people—actions should benefit harmed or vulnerable people. The status as being harmed or in a precarious position gives them buy-in for initial action and relief. Once this step is taken, turning attention to nonhuman animals and ecosystems follows. This positioning also makes sense in light of the hopeful possibility that future people will exist because they will require an environment that provides conditions for living *and* thriving. If there is an issue concerning artifacts, including but not limited to infrastructure, neighborhoods, or structures, they should come last because they are replaceable. Even though the moral order above seems absolute, it is merely suggestive. This quality allows for adjustments when a lower-ordered group's interest has a higher priority that impedes the well-being of a higher-ordered entity. This point shows that there is not so much tension between groups as there is a sense of sophistication operating in the background that requires careful parsing.

To illustrate how these notions come together to support reform, the next chapter shows what can happen if we attend to the matter above. In addition to the harms examined in Chapter 2, current food chains also impede their capacity to have intrinsic value. For instance, the ability to trade food globally that brings people together, provides billions of jobs, allows people to attain nourishment in challenging places, share culture and foods across vast distances, and merge cultures through fusion foods has a value beyond instrumentality. Having these capabilities would justify assigning intrinsic value to food chains. However, we cannot say they have this value if they also cause the myriad harms outlined in Chapter 2. In turn, this reason shows that there is an extra motivation behind making food chains ethical. If we can achieve this outcome, we can say they have intrinsic value. However, it is worth underscoring that we cannot assign it if it involves slavery and numerous other harms because nothing involving those elements deserves that label.

Further, considering that food chains' intrinsic value is made possible by numerous other areas with intrinsic value, such as mathematics and engineering principles, it is not too far-fetched to say that they have super-intrinsic value. This notion informs us about the ontological nature of food chains' intrinsic value. In turn, I provide two examples of possible saving technologies, vertical farming and a sustainable food label backed by a regulatory agency such as the United States Department of Agriculture. Although vertical farming has already shown promise for yielding technological and moral progress, a sustainable food label has monumental challenges, even though its implementation has potential. There should be plenty of motivation to see why we should strongly care about this topic with these ideas in mind.

This last point also illustrates the extent to which food chains need reform efforts, considering that current models are largely unsustainable. Working toward such an outcome will depend on piecemeal progress, requiring interdisciplinary research that depends on the most up-to-date and advanced disciplinary expertise informing wicked problems such as food chains. This notion suggests that while numerous areas of study must keep pace with agribusiness's progress, philosophy that ventures outside mainstream areas could interpret the world of food and help change it.

Epilogue
Challenges

Introduction

While the previous seven chapters illustrated the complexity of food-chain reform, the overarching issue and how I dealt with it likely warrant further investigation. This Epilogue aims to anticipate some such challenges and provide responses to them. Three issues require attention. One theoretical concern and two practical obstacles come to mind. The first concern holds that this book's thesis and suggestions amount to a disguised defense of the status quo. The second kind addresses this project's feasibility regarding the likeliness that the desired outcomes that I suggest regarding vertical agriculture will qualify as a saving technology. Instead, this technology is merely a passing trend. Sticking with and reforming established practices will pay off better in the long run, showing how addressing existing food distribution problems can deliver optimal changes realistically. The third challenge concerns the idea that government-backed sustainability labels are ridiculous, foolish, and a waste of resources, time included, even as a thought experiment. Reform efforts are better secured through passing laws geared towards global fairness, subsidies, and import–export measures. The sections below address these points in the order listed above. They show the depth that these challenges hold. Next, I confront these issues, parsing out which ones will require additional study and those that are misguided.

Challenge of Preserving the Status Quo

One could hold that any food reforms that do not aim to dismantle agribusiness amounts to nothing more than support for neoliberalism and its beneficiaries. The efforts described in this book will continue to oppress people worldwide. Instead of developing approaches that secure the global dominance of multinational food conglomerates, what is needed is a real revolution that can deliver power and food sovereignty to the people who work the fields, seas, and all food-chain links. Criticisms of the food chains that do not include these measures are distractions and clever placations that *actually subvert* real, needed reform. Such efforts, then, add to the litany of

DOI: 10.4324/9781003255505-9

offenses already committed against the people who are struggling to remove the food chains around the neck of self-determined prosperity, reconciliation, and justice.

Reply

While the challenge above seems rather severe, it has a few issues. After exposing them, the worries should settle. For instance, the first response concerns the view that any such effort that does not aim to bring down agribusiness is inherently doomed suffers from at least two shortcomings. First, this view entails that no progress can be made without some degree of criticism if it does not endeavor to accomplish a monumental, if not an impossible, task— bringing down an established institution that most of the world adopts to some degree. How is one supposed to do that? Does this mean that all such efforts are determined to fail before being presented?

This objection also impedes the idea that we should gear theoretical efforts away from relief for the people suffering unless such measures directly aim to dismantle agribusiness. This attitude will continue to harm people who are already suffering due to agribusiness. For instance, one could go as far as to say that putting unreasonable demands on reform efforts could force people working on relief issues to shelve their ideas, which would otherwise be helpful. Yet, the force of agribusiness will not make room for foundational change as well. This situation creates an impasse. In turn, this criticism is actually working to preserve agribusiness —even though it parades as if it were otherwise. Lastly, nothing in the previous chapters says that proponents who seek to bring down agribusiness cannot employ the measures described in this book to help accomplish their agenda. So, there is that.

Challenges to Vertical Farming

The first challenge concerns the idea that although agribusiness has numerous ethical issues, vertical architecture is not a one-size-fits-all solution. Any positive results that stem from such operations remain limited. There is not a significant need to implement them into food chains. Traditional practices will hold steady in the vast majority of agribusiness as the best food production and distribution measures once amended to address problem areas described previously in Chapter 2. Future technologies are likely to improve food outcomes that could inherently attend to troubling matters. Through discovering new methods of engineering food, using artificial intelligence (AI) and drone technology in fields, and improving transport, we could achieve the same if not better outcomes.

Here is the second challenge. Focusing on gimmicks such as vertical agriculture will distract researchers from advancing existing farming practices that they could improve. If technology continues to advance, it makes sense to stick with methods already working rather than radically shift the direction of farming. Unimaginable amounts of money, resources, labor, and time have gone into building and maintaining infrastructure, safety regulations, policies, and practices associated with established agribusiness. Any changes could significantly hinder these operations.

This challenge also brings another one into view. Moving attention and resources away from conventional farming practices will also hurt traditional farmers in several ways. Farming is not merely an occupation, but it is also a family business, and some family farms have been passed down for several generations. It is also a culture. These families have endured numerous hardships throughout the past several decades to survive. Farming is demanding work, to put it in simplistic terms. Nations' farmers have kept populations nourished. Showing these farmers respect should factor into future agriculture policy and reform decisions. These situations raise concerns about the interchange between farmers and the people they have fed for lengthy periods. Supporting vertical farming is not simply going to hurt established agribusiness, but it will also negatively impact farmers and their communities that depend on the economic vitality that comes with a stable industry. Even a modest market shift could harm neighborhoods, small businesses, and schools.

Although the above challenges are many, the list is non-exhaustive, suggesting that others could emerge. Still, addressing them will require significant attention. In turn, the section below does just that, giving each challenge the respect it deserves. While it would be foolish to say that I can put them to rest entirely, they should deter the pursuit of vertical technology with unyielding gusto.

Replies

The fact remains that many of the challenges above will require attention and answers beyond a philosopher's skill set due to their practical natures. Still, there is no harm in taking a few theoretical swings at them. That said, the attempts below should come with a warning and some reservations. They are suggestive, "free-style" thinking attempts only, subject to needed critique from labor scholars, historians of technology, and maverick economists.

With the above notions in mind, while the challenges listed are significant, there are plenty of ways to counter them, showing that they are not detrimental. The first challenge, that vertical farming will not solve all agribusiness' problems, is solid. The benefit of entertaining it is that it reduces the enthusiasm that borders on technological zealotry. There is a need to be rational about incorporating new technology in any field. Still, in some instances, vertical farming can complement existing food chains. This point suggests that, as parts of a larger whole, the best way to view them is as parts that improve the ethical character of specific chains or mitigate harms that come from situations such as food deserts. Further, the effects of these devices are still new, meaning that they could also have other positive impacts that could materialize, even though this aspect also entails that the opposite could arise.

The second challenge essentially holds that we should have some reservations about incorporating vertical farming into existing food chains. Doing so would divert time and resources from traditional agricultural practices, which already have had tremendous support and success. One problem is that there is no "agribusiness social contract" wherein people are obligated to purchase those products. This challenge also pushes against the norms of agribusiness. If there are no similar concessions in external areas, it seems challenging to hold that a practice without wide-scale precedent should begin with food chains.

In the same regard, the idea that other technologies could improve conventional farming practices so that there is no need to pursue sky-high production fails to consider that vertical farming could be one such technology in select cases. There is no reason why people should not support it while waiting for something better to emerge. Suppose even these artifacts' effects are marginal. In that case, the reality of increasing the availability of some fruits and vegetables in places where science shows it is advantageous should be a sought-after course, as mentioned in the last chapter. It is doubtful that vertical farming will eliminate all traditional methods, especially for some crops. Critics point out that food miles and local food will require additional assessment (McWilliams 2009). There is no reason why the pattern of this critique cannot apply to vertical and horizontal operations. This point also has implications for the next challenge: vertical farming could harm farmers, their families, and communities.

While there is a powerful argument that societies should care about people and professions, simply because new technologies could rival old ones does not entail implementing novel devices that require casting aside the humans behind their operations. Farmers are already leaving the industry for reasons that have nothing to do with vertical farming (Metcalfe 2019). Still, creating policy measures that care should pay attention to any significant challenges that arise. Moral ordering includes traditional farmers in the decision-making process at the top tiers. Such measures ensure that their needs are addressed, even though engaging in the process should be carried out on a case-by-case basis. Yet, on a practical level, successful vertical farming operations would create new jobs and opportunities, which could be available to traditional farmers.

Along the same lines, schools are now teaching about gardening in expanding ways (Earl and Thomson 2020). Select cases show how lessons about vertical farming can provide exciting ways to teach kids about agriculture and technology (Fernandez 2016). Such efforts could help draw students to this emerging technology. In the long run, these actions could help advance agribusiness, ease some burdens associated with existing work, decrease carbon emissions, water consumption, and improve the conditions and resource use linked to traditional food chains. These positive effects could also help mitigate any harm due to a shifting agricultural culture. Still, there is the possibility that farming culture would be harmed, which suggests that, at some level, affected or possibly affected farmers should be included in relevant conversations so that their interests receive attention.

Bearing these notions in mind, there is no good reason to dismiss vertical farming wholesale as a gimmick or passing trend. The idea here worth emphasizing is that once a technology shows promise, pursuing it with gusto can help reform the industry in ways that could add up as similar technologies are employed to achieve better ends. This last notion suggests that the more influence an agricultural technology can have on people, the more significant an impact it could have on the ethical nature of food chains. This last point means that a food-related technology could bring tremendous improvements if it succeeds. For this reason, this technology, along with the sustainable food label, deserves advanced study. To address this latter point, the section below moves in that direction.

The Challenge of Feasibility: The Sustainable Food Label

Despite requiring thorough assessments for social, economic, and ecological outcomes, these motivations show that entrepreneurs should study them to see how they could improve the business while securing better, more ethical outcomes for several stakeholder groups. It is beyond a doubt that implementing the requirements that a sustainable food label would demand is unreasonable at present. Still, even though it is presented as a thought experiment, one could argue that this imaginary device of ordered reason is delivered to show that it could be real in the very distant future.

The idea of developing metrics that would account for how food is grown, processed, transported, cooked in some cases, packaged, prepared, marketed, sold, delivered, and, in some instances, cooked to ensure that the stakeholder groups are treated ethically in accord with the requirements of sustainability—a term that must be defined first—pushes the bounds of the words "difficult," daunting," and "overwhelming." However, despite the tasks' monumental calling, it is also possible, worthwhile, and the right thing to do. In the worst case, even if it turns out to be impossible, we could significantly improve billions of people's lives and life on the planet.

Still, it is easy to imagine that one could call such ambitions ridiculous due to the size of the order. That criticism is fair. However, it is sad to think that food-chain problems are acceptable so that any real progress is a mirage in the distance. It gets further away as we continue reform measures that only benefit the consumers who can afford healthier food. Is it not ridiculous, perhaps in a different sense of the term, to look at the state of global food chains and accept them as they present themselves? Where is the motivation to be unflinchingly loyal to the agribusiness status quo? There is no obligation to stick to business as usual when it involves, essentially, a mind-boggling amount of harm.

Not only does pushing against agribusiness reforms preserve countless unethical practices and outcomes, but it also discounts humankind's intellectual capacities regarding science, economics, government, architecture, construction, engineering, planning, logistical analysis, and critical thinking (among others). The fact is that humanity can work towards securing the worthwhile outcomes discussed throughout this text but ignoring the naysayers should be a constant objective endured while working towards the results that will mitigate existing harm and prevent worse events from transpiring.

The radical restructuring of global food will be a gradual, slow enterprise (Epting 2016). Still, humankind can usher in a new food world, and it is acceptable if it arrives in a piecemeal fashion. Considering the magnitude of the calls for change expressed in the proceeding chapters, this point is rather humble. It acknowledges the extensive history and complicated nature of food chains. These conditions are not only deep-seated issues, but they are also woven into the fabric of most societies, extending far beyond agribusiness.

Due to this situation, I put forth the speculative need for a United States Department of Sustainability to be implemented as part of the executive branch (Epting 2021). Although such a suggestion seems extreme, the caliber and span of issues associated with wicked problems such as climate change are as well. Mitigatory efforts that can make meaningful impacts are not only forward-thinking enterprises. Humankind's continued existence of living in accustomed ways requires them. This sobering view reminds us that arriving at a highly unethical food production and distribution system took centuries, meaning that it is ambitious to think that we can change it. However, as stated above, we can.

Conclusion

The section above showed that several challenges require attention despite the ambitious nature of agribusiness reform. Even though the above sections did not identify all imaginable obstacles, the ones presented illustrate that suggested reform measures in the form of saving technologies come with the need to examine them for feasibility closely. While conceptual and practical challenges are inherent to complex enterprises, these conditions should not discourage pursuing these technologies, including vertical farming and sustainable food labels. Furthermore, present food chains have evolved over lengthy durations, meaning that reform efforts will also take ample time to create, implement, and become standard. Despite this reality, working towards a better food world will likely be a long process in a piecemeal fashion. Such a process should not be discouraging, but it should be a glimpse of the reality necessary that could lead to worthwhile, meaningful change that endures.

References

Earl, L., & Thomson, P. (2020). *Why Garden in Schools?*New York: Routledge.

Epting, S. (2016). Participatory budgeting and vertical agriculture: A thought experiment in food system reform. *Journal of Agricultural and Environmental Ethics*, 29 (5), 737–748.

Epting, S. (2021). The United States Department of Sustainability: Its purpose and moral underpinnings. *Sustainability and Climate Change*, 14(2), 84–91.

Fernandez, D. (2016). Oak Cliff Students use vertical farming lesson to help community. WFAA 8 ABC, April 21. https://www.wfaa.com/article/news/education/oak-cliff-students-use-vertical-farming-lesson-to-help-community/287-148022886.

Metcalfe, R. (2019). *Food Routes: Growing bananas in Iceland and other tales from the logistics of eating*. Cambridge, MA: MIT Press.

McWilliams, J. (2009). *Just food: Where locavores get it wrong and how we can truly eat responsibly*. New York: Little, Brown and Company.

Index